HACKS FOR
ALEXA

HACKS FOR
ALEXA

AN UNOFFICIAL GUIDE TO SETTINGS, LINKING DEVICES, REMINDERS, SHOPPING, VIDEO, MUSIC, SPORTS, AND MORE

JOSEPH MORAN

Racehorse Publishing

Racehorse Publishing books may be purchased in bulk at special discounts for sales promotion, corporate gifts, fund-raising, or educational purposes. Special editions can also be created to specifications. For details, contact the Special Sales Department, Skyhorse Publishing, 307 West 36th Street, 11th Floor, New York, NY 10018 or info@skyhorsepublishing.com.

Racehorse Publishing™ is a pending trademark of Skyhorse Publishing, Inc.®, a Delaware corporation.

Visit our website at www.skyhorsepublishing.com.

10 9 8 7 6 5 4 3 2 1

Library of Congress Control Number: 2019912318

Cover artwork credit: Amazon.com
Cover design credit: Brian Peterson

Print ISBN: 978-1-63158-530-2
Ebook ISBN: 978-1-63158-533-3

Printed in the United States of America

CONTENTS

INTRODUCTION

"Is it like Google that you can talk to?" This question was posed to me not all that long ago by client who wanted to know what Alexa, Amazon's artificial intelligence virtual assistant, was all about before deciding whether it was something worth looking into.

Although my client's impression of Alexa wasn't exactly incorrect, it was certainly incomplete, because Alexa can be so much more than simply a voice-activated search engine. Plus, for the record, you can talk to Google from any PC or mobile device (note the microphone icon on the right side of Google's search box) and Google offers its own virtual assistant called Google Assistant.

I've also had numerous other clients, not to mention various friends, family, and neighbors, frequently pepper me with similar questions about what, exactly, Alexa is and what it can do for them.

Hacks for Alexa aims to answer those questions.

If you've had any exposure to Alexa before, in person or even just from a TV commercial, you're probably aware that Alexa can answer questions such as "What's the temperature outside?" "What's a synonym for happy?" or "How many teaspoons are in a tablespoon?" So, in fact, on a basic level, Alexa is a lot like having someone who knows a lot of stuff—or at least can Google things for you very quickly—at your beck and call.

You may or may not also know that Alexa can fulfill all kinds of peculiar and/or whimsical requests—for example, to tell a joke, roll a 20-sided die (or a 20,000-sided one for that matter), or make farting noises. But while asking Alexa to provide general information or perform parlor tricks can certainly be useful and entertaining, using Alexa this way only scratches the surface of its capabilities.

Hacks for Alexa doesn't set out to be a comprehensive guide to everything you can possibly do with Alexa. Alexa's abilities are vast and constantly growing, so no book could realistically hope to accomplish that. But precisely because Alexa can do so many things, it's not always clear what it's capable of and how to best take advantage of it. That's where this book comes in.

The goal of this book is to serve as a handy reference to enhancing your lifestyle with Alexa. We'll spend some time exploring how Alexa works behind the scenes and reviewing the different kinds of Alexa-enabled devices that are available. (Alexa-enabled devices are those that you can talk to Alexa through.) Then we'll explore various handy, and often overlooked, ways Alexa can help you organize, simplify, automate, and enhance your daily life.

In this book, you won't learn how to hack Alexa, but you will learn ways to use Alexa to hack and improve your lifestyle. These include how to:

- Keep organized via lists, timers, reminders, and alarms
- Check your email and manage your calendar
- Make and receive voice (or video) calls with other Alexa users, plus call landline and mobile phones
- Use Alexa as a home intercom
- Find out what traffic is like on your commute
- Get detailed local weather information
- Play music in any room of your home (or in every room)
- Make Amazon purchases using your voice and let you know when shipments are on the way or have been delivered
- Use the foundation of a modern smart home, providing voice control of lights and other connected devices
- Have Alexa listen for trouble or potential intruders when you're away from home
- Make Alexa kid-friendly
- Manage your privacy
- And much more

Here are a couple of important things to remember as you read this book:

Although Alexa has an ostensibly female name and female voice, we're not going to refer to Alexa as "she," but rather as "it." Although it can be easy to forget sometimes, Alexa, of course, isn't a real person or female, but a genderless artificial intelligence.

Alexa is constantly receiving new and improved features, so there's a chance that certain things may look or work a bit differently than we describe in this book because of changes made to Alexa since the book went to press.

Thanks very much for buying *Hacks for Alexa*. I hope you find it useful.

1
ALEXA BASICS

Alexa, Amazon's voice-operated virtual assistant, which uses artificial intelligence to respond to questions and commands, is useful not simply because of the things it can do; much of what Alexa can do can often be done from a PC or a smartphone app. Alexa's utility and appeal lies mostly in the fact that you can interact with it almost entirely by voice, and do so conversationally, much as you would with a real human being. The ability to get things done by voice rather than by typing and tapping can save a lot of time and effort (not to mention keep your hands free for other tasks), which explains Alexa's significant and growing popularity.

At the beginning of 2019, Amazon reported that the Echo Dot—one of several Alexa-enabled devices the company makes—was its best-selling product worldwide in 2018 (this includes products Amazon sells but doesn't make itself). So, if you've recently become an Alexa user or are thinking about becoming one, you've got a lot of company.

HOW ALEXA WORKS

You don't need to know exactly how an automobile's engine works to drive one, and similarly, you don't necessarily need to know the details of how Alexa works in order to use it. But when you're planning to add an artificially intelligent virtual assistant to your daily life, it's helpful to have a basic understanding of what Alexa is and what goes on under the hood.

To best understand what Alexa is, it helps to understand what it's not, and it's not simply a device that you buy—at least not entirely.

If you're of a certain age, you may remember the seventies-era *Charlie's Angels* TV show (and subsequent movies), in which detective agency owner

1

Charlie was never there in person. Rather, he ran things from some distant locale and so always communicated with his "angels" (and they with him) via a speakerphone box sitting on a desk.

Alexa works a lot like Charlie's speakerphone box in that you're talking to someone (or in this case, something) far away, and that Alexa-enabled device sitting on your end table or kitchen counter isn't really Alexa, but rather a conduit to Alexa.

We'll discuss the specifics of Alexa-enabled devices in more detail a bit later in this chapter, but first let's take a quick look at what really powers Alexa.

THE ALEXA VOICE SERVICE

The Alexa Voice Service (AVS) is the brains behind Alexa, where most of the real heavy lifting is done. It's in the AVS, a cloud computing service located in a far-off data center and linked to your Alexa-enabled device via an Internet connection, where Alexa puts various components of artificial intelligence to work. These include Automatic Speech Recognition and Natural Language Understanding which process what you say, try to understand what you meant (usually, though not always, successfully), and then decide what response to give and/or action to take.

Keeping most of Alexa's smarts in the cloud rather than on the device has pros and cons. The major benefit of this approach is that, thanks to continuous development work by Amazon, Alexa is always evolving. It frequently receives new and improved capabilities that you can immediately take advantage of without having to purchase new hardware. As a result, the Alexa-enabled devices you talk to don't need a lot of computing resources such as processing power or storage, allowing them to be simpler and relatively inexpensive. Since Alexa's capabilities are largely the same regardless of whether you bought your device last year or just it took out of the box yesterday, you get a degree of protection against hardware obsolescence.

But there's also a downside to having Alexa's intelligence primarily in the cloud, and it's that most Alexa-enabled devices can do little to nothing without a connection to the AVS. (There are a few exceptions, which we'll talk about later.) This means if your Internet connection goes out, for all intents and purposes, Alexa is out of service too.

Even when your Internet connection is functioning normally, it's still possible to lose access to Alexa due to a problem with the AVS itself. Like any service that runs in the cloud, AVS is susceptible to outages. These outages are generally infrequent and short-lived, but they do happen. On several occasions while writing this book, Alexa was unable to answer questions or perform tasks, even though my Internet connection was up and running. And on one recent occasion the AVS outage was widespread and prolonged enough to have been widely publicized on numerous Internet media outlets.

Anatomy of an Alexa Interaction

Let's examine a somewhat simplified sequence of events that takes place when you have an interaction with Alexa:

- You say Alexa's name within range of an Alexa-enabled device's microphone, which alerts it to begin recording audio and send it to the AVS. (The word that gets Alexa to start paying attention to you is known as the wake word, and it can be changed to something other than "Alexa"—more on this in chapter two.)
- The audio recording is received by Alexa's Automatic Speech Recognition (ASR), which processes the sound of your voice and converts it into one or more strings of text containing identifiable words.
- Alexa uses Natural Language Understanding (NLU) to analyze the spoken words and the order in which they were used to understand their meaning and thus the user's likely intent.
- Alexa connects to either internal or external resources to obtain the information necessary to satisfy the request.
- The information needed for the response is recorded into an audio file, which is then streamed back to your device for playback. (The response may be an answer to a question, but it can also be a follow-up question.)

ALEXA-ENABLED DEVICES

As mentioned previously, a device you talk to Alexa through is known as an Alexa-Enabled device. Naturally, Amazon manufactures Alexa-Enabled devices—the company's "Echo" line includes about a half-dozen different models. (There are also Alexa-Enabled devices available from third-party manufacturers, which we'll discuss shortly.)

All Echo devices have the same core purpose, which is to serve as an input/output device for Alexa. To that end every Echo device has multiple built-in microphones so they can hear from any direction (and often at a considerable distance). Most Echo devices also have a built-in speaker, though a few don't because they're intended to connect to external speakers. Some Echo devices even have touch screens, enabling video calling and allowing Alexa to show as well as tell you things.

Aside from the presence of a screen or a speaker, the other major differentiator between Echo devices is audio quality. Some Echo devices provide better audio quality than others due to larger internal speakers and/or more of them, which can be an important consideration if you plan to use Alexa to play music frequently. (See more on this in chapter six.)

A common concern when buying technology devices is the fear that not long after you buy one, a new-and-improved version comes out that costs less and does more. Amazon refreshes its Echo devices roughly once a year, and newer Echo devices do generally offer better aesthetics and better-sounding audio at a lower—or at least the same—cost. But remember that Alexa's smarts are primarily in the cloud, not in the device, so in most cases you'll be able to take advantage of new features without having to buy new hardware.

Here's a rundown of Amazon's current line of Echo devices available as of late 2019:

Echo Devices with Built-In Speakers
Echo Dot
The Echo Dot (Figure 1-1), currently in its 3rd generation, is the mainstream, entry-level device in the Echo line. It's available in Charcoal, Heather Grey, and Sandstone fabric finishes (read: black, gray, and white) and typically sells for $50, though it can often be found for even less during periodic Amazon promotions and/or when you buy a pair of them. Although earlier generations of the Echo Dot suffered from a tinny, hollow-sounding audio, the 3rd generation model's single 1.6-inch speaker offers dramatically improved sound.

Figure 1-1: The Echo Dot (image credit: Amazon)

Echo

The Echo (Figure 1-2), currently in its 2nd generation, has a similar footprint to the Echo Dot but is much taller (5.8 inches versus 1.7 for the Dot) and sells for $100, though like the Dot it's frequently discounted. The Echo offers improved audio quality and improved bass over the Echo Dot because of its pair of internal speakers (a 2.5-inch woofer and 0.6-inch tweeter) as well as Dolby audio processing. The Echo is available in the same three fabric finishes as the Echo Dot, but its outer shell can be removed and replaced with a variety of decorative shells ($30 each) in wood or fabric styles.

Figure 1-2: The Echo (image credit: Amazon)

Echo Plus

The Echo Plus (Figure 1-3), also in its 2nd generation, normally sells for $165 and comes in the same three fabric finishes as the Echo Dot and Echo (no removable shells here, though). The Echo Plus further improves on the Echo's audio capabilities with a larger Dolby-tuned woofer and tweeter (3 and 0.8 inches, respectively). The Echo Plus also contains a built-in Zigbee hub for smart home automation and is the only Echo device with a room temperature sensor. (For more on Zigbee and the temperature sensor, see chapter eight.)

Figure 1-3: The Echo Plus (image credit: Amazon)

Echo Auto

As you might have guessed from the name, the Echo Auto (Figure 1-4) is designed to let you use Alexa behind the wheel. The $50 device mounts to a car's dashboard, gets its power from a USB or cigarette lighter port, connects to your car's speakers via Bluetooth or an AUX input (3.5mm cable), and connects to the Internet via your smartphone. As of publication, the Echo Auto is available for purchase on an invitation-only basis.

Figure 1-4: The Echo Auto (image credit: Amazon)

Echo Input

At $20, the Echo Input (Figure 1-5) is Amazon's least-expensive Echo device. Available in either black or white and somewhat resembling a thick drink coaster, it lacks an internal speaker so the best way to think about the Echo Input is as an Echo Dot with BYOS (Bring Your Own Speaker).

Figure 1-5: The Echo Input (image credit: Amazon)

Echo Devices with Screens

Echo Spot

The Echo Spot (Figure 1-6) is the smallest of three Echo devices equipped with a touch screen. The $130 Echo Spot is compact and spherical, has a single 2-watt speaker, and measures 4.1 x 3.8 inches. It sports a circular 2.5-inch screen as well as a front-facing camera for video calling.

Figure 1-6: The Echo Spot (image credit: Amazon)

Echo Show

Think of the Echo Show as the Echo Spot's much larger, more powerful, and pricier sibling. The $230 Echo Show (Figure 1-7) measures 6.9 x 9.7 x 4.2, sports a 10.1-inch HD screen, and includes dual ten-watt speakers with Dolby audio processing. Like the Echo Dot, the Echo Show has a front-facing camera, and like the Echo Plus, it has a built-in Zigbee smart home hub.

Figure 1-7: The Echo Show (image credit: Amazon)

Figure 1-8: The Echo Show 5 (image credit: Amazon)

Echo Show 5

The Echo Show 5 (Figure 1-8) is a smaller-scale version of the Echo Show, and at $90 is currently the least-expensive Echo device equipped with a screen. Compared to the Echo Show, the 3.4 x 5.8 x 2.9 Echo Show 5 has a smaller 5.5-inch (and non-HD) screen, a single 4-watt speaker, and lacks a Zigbee hub. One thing the Echo Show 5 has that its larger sibling lacks is a shutter switch to physically cover the front-facing camera lens.

Certain recently released Amazon Fire tablets support "Show Mode", which give the tablet most of the same capabilities of Echo Show/Spot devices. For more information about which Fire models support show mode and how to use it, see www.amazon.com/gp/help/customer/display.html?nodeId=G3JEMFDNH3FPB4XS

THIRD-PARTY DEVICES WITH ALEXA "BUILT-IN"

In addition to Amazon's own line of Echo devices, there are numerous products from manufacturers other than Amazon that advertise themselves as having "Alexa Built-in," a moniker which means that the device includes a microphone, a speaker, and the ability to connect the Internet and Amazon's Alexa Voice Service (AVS).

There are a wide range of Alexa Built-in products available, and often, though not always, their ability to talk to Alexa is secondary to the primary function of the device. Examples of such Alexa Built-in devices include speakers, soundbars, headphones, earbuds, tablets, Wi-Fi access points, smartwatches, thermostats, smoke detectors, and more.

> For a current list of Alexa Built-In products, see https://www.amazon .com/b/ref=EchoCP_avs_tile_text?node=15443147011

What's important to know about Alexa Built-in products is that they don't necessarily support all the same features of Alexa that an Amazon Echo does. This is because Amazon doesn't make every Alexa feature available to third parties, and sometimes the third-party chooses not to implement a feature in their product.

The Sonos One, for example, an audiophile-level speaker with Alexa Built-in, doesn't support Alexa's communication features (which we cover in chapter five). Similarly, the Eufy Genie, a compact device like the Echo Dot that sells for less than half the Dot's typical price, doesn't offer a way to change the wake word (which we discuss in chapter three).

Other Alexa Built-in devices may have different limitations, so while choosing an Alexa Built-in device over an Amazon Echo can make sense in some situations, read the product's description carefully before buying so you know which Alexa features, if any, aren't supported.

> Alexa Built-in devices are not to be confused with devices labeled "Works with Alexa," which denotes devices that are certified to work compatibly with Alexa (i.e., Alexa can control them). In a nutshell, *Alexa Built-in* means a device lets *you* talk to Alexa, while *Works with Alexa* means that Alexa can talk to the device.

ECHO LIGHT RINGS

Although Alexa is voice-oriented, the Echo, Echo Dot, and Echo Plus feature a light ring that pulses and spins in various colors (blue, cyan, orange, red, green, yellow, purple, and white) to give a visual status of the device or what Alexa is currently doing. Echo devices with a screen don't have light rings and instead display their status as a horizontal color bar at the bottom of the screen.

Here are the various Echo light ring states and what they mean:

- **Solid blue with spinning cyan**
  The device is starting up. If you didn't just plug the device in, it probably just had its power restored or is restarting after a software update.
- **Spinning orange clockwise (during initial device setup)**
  Device is connecting to Wi-Fi.
- **Pulsing purple (during initial device setup)**
  There was a problem connecting to the Wi-Fi network.
- **Solid blue with a segment of cyan**
  Alexa is listening to and processing your request. The cyan segment indicates the direction in which speech is being detected and will move to follow the voice of the speaker.
- **Alternating blue and cyan**
  Alexa is responding to your request.
- **Pulsing yellow**
  A message or notification awaits.
- **Pulsing green**
  Indicates an incoming call or Drop In from another Alexa user.
- **Spinning green**
  Indicates an active (in progress) call or Drop In.
- **One purple flash (following an interaction with Alexa)**
  Do Not Disturb mode is enabled for the device.

For more on Alexa's communication features such as calls, messages, and Drop In, as well as Do Not Disturb mode, see chapter five.

- **Solid red**
  The device's microphone has been turned off.
- **Solid white**
  Indicates the volume level of the device (only appears while you are adjusting the volume).
- **Spinning white**
  Alexa Guard is set to Away Mode.

For more on Alexa Guard, see chapter eight.

By the way, Echo devices have no power light (or power switch, for that matter), so a properly functioning Echo device looks no different from one that's not even plugged into the wall or one that is plugged in and powered up but isn't working because it doesn't have an Internet connection.

A quick and easy way to check whether a dark Echo device is both plugged in and connected to the Internet without having to physically examine it is to ask Alexa a simple question, such as *"Alexa, what time is sunset today?"* If you don't get a response, it's a safe bet the Echo device isn't plugged in. If you get a response of, "Sorry, I'm having trouble understanding right now. Please try a little later," or something similar, Alexa can't get the answer because your Internet connection isn't working (or else the AVS itself is having problems).

THE ALEXA APP

You might be thinking, *If I control Alexa with my voice, why do I need an app?* Fair question, but it turns out that the Alexa app—available on both Android and Apple devices (and included with Amazon's own Fire tablets)—is an integral part of the Alexa experience, so you'll always want to have it close at hand.

You'll need the Alexa app to set up a new Echo device for the first time, but even after that the Alexa app is often the best way—and sometimes the only way—to enable or configure certain Alexa features.

Moreover, you can talk to Alexa via the Alexa app, so it's essentially an Alexa-enabled device in your pocket you can use when you're away from home, or at home but not in the vicinity of an Echo device (out in the yard, for example). There are, however, certain things you can do with an Echo device that you can't do with the Alexa app—for example, talk to Alexa without having to tap a microphone button first.

The upshot is that you should always have the Alexa app handy on your smartphone (we'll assume you have a smartphone if you're interested in an artificial intelligence voice assistant like Alexa). Also, even if you don't update your other smartphone apps automatically, be sure to check for and download new versions of the Alexa app often—Amazon updates the app frequently and these updates are often required to access new features as well as fix bugs in existing features.

If you don't have a smartphone or tablet to run the Alexa app on (or don't have it handy), you can access many of the same configuration settings from a PC or Mac via a web browser by pointing it to alexa.amazon.com. Compared to the Alexa app, some of the website menus are organized a bit differently, and some features are only configurable from the app. Also, you need the Alexa app to set up new Echo devices, as you can't do it from the website.

We'll be using the Alexa app (Figure 1-9) extensively throughout this book, so let's take a quick tour:

The Menu button (the stacked lines at the upper left) provides access to the main menu, where you can access various Alexa features, install and manage skills, and change settings.

There's a row of five icons along the bottom of the app. Starting from the left:

Home: Displays the Home screen, which always shows the current date and weather forecast along with a scrollable list of cards that show your recent interactions and activity with Alexa, with the most recent on top. These cards are interspersed with Alexa's "Things to Try" cards, which suggest features, skills, etc.

Figure 1-9: The Alexa app Home screen

Communicate: Here's where you can use Alexa's communications features, including calling, messaging, Drop In, and announcements.

For more on Alexa's communications features, see chapter five.

Alexa: Tapping this Alexa logo in the bottom central of the app will let you speak commands and questions to Alexa through the device, provided you've given the app permission to access the device's microphone. The first time you try to use this feature, the app will either ask you to confirm you want to allow it to use the microphone, or automatically take you to the setting where you need to give the app permission to access the microphone (this will vary depending on the type of device you're running the app on).

Play: Here you'll be able to review and replay any recently accessed media such as music or audiobooks, as well as browse audio content from Amazon Music and other Alexa-compatible services.

Devices: This is where you can view and configure your Alexa-enabled and Alexa-compatible devices, as well as create groups for speakers and smart home devices, such as lights.

For more on Alexa's music features, see chapter six, and for more on smart home devices, see chapter eight.

UNDERSTANDING ALEXA SKILLS

Amazon has built a lot of features directly into Alexa and adds new features all the time. But just like you and I, Alexa wasn't born knowing how to do everything that might be asked of it. When there's something Alexa can't do "out of the box," that's where skills come in.

Alexa skills are the equivalent of apps on a smartphone—they give Alexa new abilities, typically by allowing access to and interactions with external services, devices, or information sources.

For example, Roku offers a skill that lets Alexa control your Roku streaming device or Roku-equipped TV via voice commands. Other skills give Alexa the ability to do things such as host a version of the TV game show *Jeopardy* (as well as various other games), tell you how to mix a cocktail, or play relaxing sounds such as falling rain or ocean waves, to name just a few.

As it happens, there are tens of thousands of Alexa skills available (in early 2019, Amazon reported over 80,000 Alexa skills exist worldwide). Alexa skills are always free, but just like smartphone apps, many skills offer paid upgrades—additional features or content—in exchange for a one-time payment, or increasingly, an ongoing monthly subscription. Unlike smartphone apps, you don't download or install skills onto specific devices, but rather attach or "enable" them on your Alexa account, which makes them accessible on any of your Alexa devices.

Most skills are ready to use as soon as you enable them, but skills that need to interact with your account on another services—say, one that lets Alexa play music from your Pandora account—will ask you to link the skill by providing access permission and your login information for that account.

Finding and Enabling Skills

There are several different ways to make use of Alexa skills. The most straightforward way is via the Alexa app. Tap **Menu** > **Skills & Games** and search for skills by name or keyword or browse them by category (Figure 1-10). When you find one you want, tap the **Enable to Use** button to activate it, and note the "Start by Saying" section, which gives examples of different things you can say to use the skill.

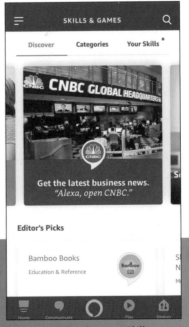

You can also browse, search, and enable Alexa skills from a web browser by visiting www.amazon.com/alexa-skills/b?ie=UT-F8&node=13727921011.

Figure 1-10: The Alexa app Skills screen

Once you've enabled a skill, you can tell Alexa to open it via a voice command; just say, *"Alexa, open {skill name}"* to use the skill, though in some cases, simply saying, *"Alexa, {skill name}"* accomplishes the same thing.

Letting Alexa Handle Skill Selection

It wasn't too long ago that you first needed to find and enable skills from the Alexa app as described above, then remember the skill's name (also known as the "invocation name") and explicitly ask Alexa to open the skill by name before you could use it. But as we mentioned earlier, Alexa is constantly receiving behind-the-scenes improvements to get smarter, and one of those improvements makes it easier to use some skills without having to enable them first, or even know what specific skill you want to use.

In many cases, when you ask Alexa to do something that isn't in its standard repertoire, Alexa can automatically suggest, enable, and open a relevant skill to fulfill that request. For example, if you say, *"Alexa, I need a ride,"* and you've already enabled the skill from either Lyft or Uber, Alexa will automatically open that skill. But if you haven't previously enabled one of these ride-sharing skills, Alexa will ask you whether you want to try Lyft or Uber (one at a time, and not necessarily in that order) and then enable and open that skill. In this example, Alexa will direct you to the Alexa app to link the skill to your account for that service, but skills that don't require account linking would be ready to use.

Note that Alexa can only tell you about and automatically enable those skills that have been designed to support this feature. As skill developers update their skills over time, more and more skills will be accessible this way, but depending on the skill you still may have to enable and use it via the old-fashioned method described earlier.

Managing Alexa Skills

Regardless of whether you enable a skill through the Alexa app, the Amazon website, or a voice command, you may want to periodically review the skills you've enabled, especially if you enable a lot of them via voice command.

You can do this from the Alexa app by choosing **Menu > Skill & Games > Your Skills** (Figure 1-11). Here you'll see how many skills you've enabled along with how many need your attention—usually because they can be linked to an outside account but aren't. (The third category, "Blueprint," refers to skills you've created yourself. See more about this in chapter ten.)

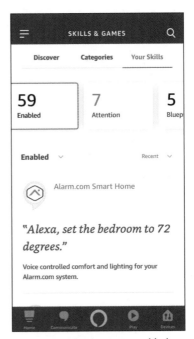

Figure 1-11: Viewing your enabled skills in the Alexa app.

2
CONTROL AND RESPONSE

In this chapter we're going to explore various ways you can tweak settings to customize your basic interactions with Alexa and how Alexa responds to you. Among the things we'll cover in this chapter are changing Alexa's wake word, naming devices, controlling volume and the microphone, changing Alexa's language or accent, and getting Alexa to recognize different voices.

ACCOUNT SETTINGS VERSUS DEVICE SETTINGS

Before we go any further, it's important to note that Alexa's numerous features and configuration settings fall into two basic categories—device-specific and account-specific. Device-specific features are those that apply only to a specific Echo or other Alexa-enabled device. For example, when you change Alexa's wake word, or set an alarm, reminder, or timer, those things only apply to a single device.

In contrast, there are other features and settings that apply to your entire Alexa account, and thus apply no matter what device you're interacting with Alexa on. Examples of account-specific include things like lists, Whisper Mode, and Voice Purchasing.

USING MULTIPLE ECHO DEVICES

It only takes a single Echo device to bring Alexa's capabilities into your home but having multiple Echo devices not only allows you to speak to Alexa in more places, it opens up a world of additional features including the ability to use Alexa's whole-home intercom or public address system, or for multi-room music. (Learn more about these features in chapters five and six.)

As mentioned earlier in this chapter, every Echo device contains multiple microphones designed to pick up your voice from the surrounding area, even at a considerable distance. So, when you talk to Alexa with multiple Echo devices listening nearby, which one answers? Alexa uses a feature called ESP (Echo Spatial Perception) to determine which device is closest to you and delivers the response from that device.

ESP technology isn't perfect, and every so often you may find that instead of receiving a response from the nearest Echo device, you get it from one down the hall or even in the next room. This isn't usually a problem, though it may be annoying for device-specific features such as alarms and timers. (You wouldn't want a timer you set in the kitchen go off in the bedroom, for example.)

NAMING YOUR ECHO DEVICES

Not to be confused with the wake word (which we'll talk about next), every Echo device has a name of its own. When you set up an Echo device for the first time via the Alexa app, it gets a generic name such as Joseph's Echo or Joseph's Echo Dot unless you specifically choose to give it a customized name. Add another Echo device later, and it will be automatically be named something like Joseph's 2nd Echo Dot, and so on.

Nondescript names can become a problem when you have more than one Echo device, because it makes it hard to know which device you're dealing with if you want to change a device-specific setting. It also causes confusion if you want to use communication features between a pair of Echo devices or send music to a specific Echo device (which we describe in chapters five and six, respectively).

For this reason, if you have more than one Echo device at home, it's a

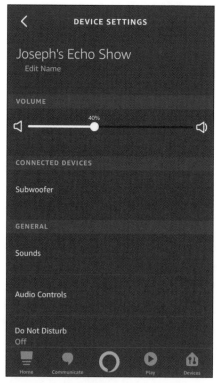

Figure 2-1: Tap Edit Name to change the name of an Echo device.

good idea to give each one a descriptive name indicating where it's located, such as Kitchen, Master Bedroom, or Underground Lair. To change an Echo device's name, from the Alexa app, tap **Devices** > **Echo & Alexa**, and then the name of the Echo device you want to rename. Then tap **Edit Name** under the device's current name to rename it (Figure 2-1).

By the way, if you're trying to rename several generically named Echo devices and don't know which is which, you can just ask Alexa at a particular Echo device. Say, *"Alexa, what's your device name?"* and Alexa will respond with "It's {device name}," so you'll know which one to look for in the Alexa app menu. It's important to use the word "device" in this command—if you only ask, "What's your name?" Alexa will respond that her name is Alexa.

As of this writing, it's not possible to change an Echo device's name using your voice—only with the app. If you say, *"Alexa, change your device name,"* Alexa will offer you options to change the wake word instead.

CHANGING THE WAKE WORD

As we discussed in the last chapter, the wake word is what triggers Alexa to know that whatever you say next is something it needs to process and respond to. But there may be a reason you want Alexa to answer to something other than the name "Alexa," one of which might be that Alexa is the name of someone that lives in or frequently visits your household.

You can change the wake word for any reason you want to, though not to any word you want. There are three other wake word options available—"Amazon," "Echo," or "Computer." (That last one is a nod to Star Trek fans, allowing them talk to Alexa the same way Enterprise crew members addressed the ship's computer. Evidently no one back then considered that a computer would have a proper name.)

To change the wake word for an Echo device, just go to it and say, *"Alexa, change your wake word to {wake word}."* The Echo light ring will glow orange for a moment and Alexa will answer: "OK. You can call me {the new wake word} on this device in a few seconds." By the time Alexa's finished saying that, you can start using the new wake word.

The important thing to remember about changing the wake word is that it's device-specific, so if you have several Echo devices and change the wake word on only one of them, the others will still respond to "Alexa." To change the wake word for multiple Echo devices without having to visit them all in person, use the Alexa app. Tap **Devices > Echo & Alexa**, the specific Echo device you want, then scroll down to **Wake Word** and select your new wake word from the list (Figure 2-2).

As if to emphasize that Alexa is an entity apart from an Echo device and its wake word, if you change a wake word to, say, "Computer," and then ask, *"Computer, what's your name?"* the response will still be, "My name is Alexa."

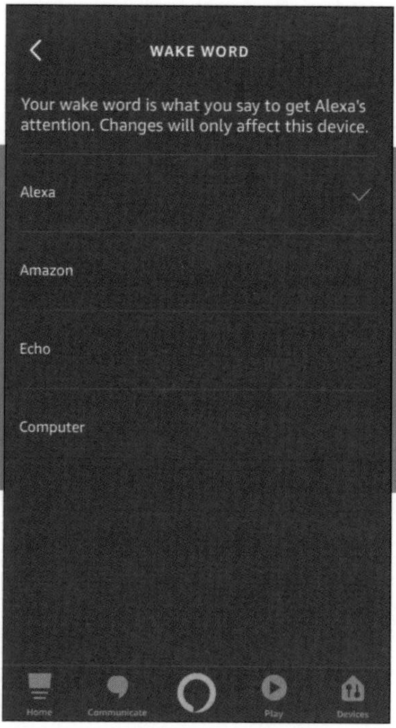

Figure 2-2: Changing an Echo device's wake word from the Alexa app.

ENABLE FOLLOW-UP MODE (TO SAY THE WAKE WORD LESS OFTEN)

Now that we've shown you how to change the wake word, let's look at a way to say that wake word less often.

The typical way to interact with Alexa is to first say the wake word to get Alexa's attention, then ask your question or speak your command, and finally let Alexa reply and/or take whatever action you requested. After that, Alexa tunes out until you say the wake word again. But when you want to ask Alexa multiple questions or give multiple commands, going through the *wake word-question-response* sequence repeatedly can be cumbersome and time-consuming. (Not to mention a bit unnatural—imagine having to speak to a real person that way.)

But with a feature called Follow-up Mode, you can speak to Alexa a bit more like you would an actual human being. To turn on Follow-up Mode, open the Alexa app, tap **Devices > Echo & Alexa**. Then choose a device, scroll down to **Follow-up Mode**, and turn it on (Figure 2-3).

As with the wake word, you need to turn on Follow-up Mode for each device individually.

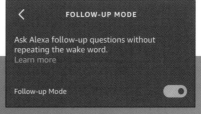

Figure 2-3: Turn on Follow-up Mode to give multiple commands after one wake word.

With Follow-up Mode enabled, after you ask Alexa a question and receive a response, you'll notice that the light ring remains blue for about five seconds after Alexa finishes speaking. This is your cue that Alexa is still listening, allowing you to make additional requests without using the wake word again.

Note that each new input doesn't necessarily have to be related to what came before it. For example, Follow-up Mode allows you to ask Alexa a series of related questions, such as:

"Alexa, how far away is Key West?"
"What's the temperature there?"
"How long will it take to drive there?"

Or you can Ask Alexa completely unrelated questions, such as:

"Alexa, when was George Washington born?"
"What's the square root of 4,096?"

There doesn't seem to be a limit to how many successive questions Alexa can answer in Follow-up Mode. Just remember to ask your subsequent questions within the five second window while the light ring remains lit. Also, note that follow-up doesn't work when Alexa is playing audio.

You can end a Follow-up Mode session before the five seconds is up by saying "Thank you," "Stop," or "Cancel."

CONTROLLING VOLUME

You might think controlling the volume on an Echo device is a relatively straightforward task, and it can be, but it can also be a little confusing because the volume controls are somewhat different depending on how you access them.

Here are the three options for controlling Alexa's volume on an Echo device and variations for each.

By Voice

You can adjust Alexa's volume by saying, *"Alexa, set volume to {a number between one and ten}."* You can ask Alexa to adjust volume by half-steps too . . . well, you can ask, but Alexa won't do it. Ask Alexa to set volume to six and a half and it'll ignore the "half" and just set it to six. (Yet if you ask Alexa to set volume to six and three-quarters, Alexa tells you it can only set volume in whole numbers. Weird.)

In the Alexa App

To adjust an Echo device's volume from the Alexa app, tap **Devices > Echo & Alexa**, then choose the specific Echo device you want to adjust. Here you'll see a slider control (Figure 2-4) that you can set not on scale of one to ten, but between one and one hundred in one percent increments.

Figure 2-4: Control an Echo device's volume from the app while it's playing.

You'll only see the volume slider when the Echo is playing audio, so to change the volume of an Echo that's not playing, use voice commands or the physical volume buttons.

Using Physical Volume buttons

Every Echo device has a pair of volume buttons labeled "+" and "-" (Figure 2-5) which are probably the least convenient way to adjust Alexa's volume, but you'll still probably wind up using them sooner or later. So, if you can set the volume from one to ten by voice, and from one to a one hundred in the Alexa app, how many volume button presses does it take to go from zero to maximum volume?

Ten? Nope. One hundred? Nope. Turns out it takes thirty presses of the volume button to go from no volume to full volume. Math folks may notice that 30 is not a divisor of 100 (unlike 20 or 25), which means each press of the volume button raises or lowers the volume by either three or four percent.

What's the takeaway from all this? For blunt but convenient control over volume, use voice commands. For less convenient but finer volume control,

Figure 2-5: Physical buttons on an Echo Dot. (image credit: Amazon)

use the volume buttons or the app, depending on whether you want to get up from your chair or not.

Muting the Speaker

To mute Alexa's speaker while it's playing audio, just say, *"Alexa, mute."* Alternately, you can use the Echo device buttons or the Alexa app's volume slider as previously mentioned to set the volume to down to zero. Regardless of how you mute it, a small segment of the Echo's light ring next to the volume down button will blink white three times to indicate the device is muted. The ring then goes dark, so there's ongoing visual indication that an Echo is muted.

It's important to note that mute on Alexa works just like muting on any other device—when you mute, whatever you were listening to continues to play—something to keep in mind if you're listening to material such an audiobook or a podcast and don't want to miss anything. To temporarily stop the playback of audio rather than just silencing the audio, say, *"Alexa, pause."*

Setting the Equalizer

In addition to volume, Echo devices also provide basic equalizer settings for Bass, Midrange, and Treble which you can set in a range between minus 6 and plus 6 (Figure 2-6). You can view and adjust these settings from the Alexa app by tapping **Devices > Echo & Alexa**, the Echo device you want, then **Audio Controls** (which you'll find under the **General** category). You can also ask Alexa to set the equalizer via voice command, for example, say, *"Alexa, set bass to minus three."*

DISABLING AN ECHO'S MICROPHONE

If there are times you want privacy and don't want Alexa to listen to you or respond to the wake word, you can disable the Echo's microphone. (For more detail on how Alexa listens to, records, and retains what it hears, see chapter nine.) You can't disable an Echo's mic by voice command for reasons which we'll discuss in a moment, so the only way to do so is by pressing a physical button on the device itself.

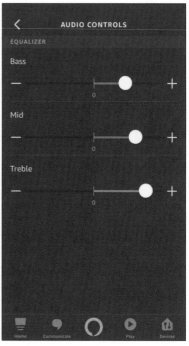

Figure 2-6: The Echo's audio equalizer settings.

On some Echo devices the button is labeled with a microphone with a line through it, while others use what appears to be an "O" with a line through it (Figure 2-5). But either way, when you press this button, the button itself and the Echo's light ring will continuously glow solid red indicating that Alexa is no longer listening through that device.

If it seems inconvenient to have to get up and press a button to disable the microphone rather than simply saying something like "Alexa, disable microphone," it is, but it's done for a good reason—namely to ensure the microphone can't be turned back on remotely and/or without your knowledge.

When you press the microphone button on an Echo device, you're physically disconnecting electric power to its microphone. Echo devices are designed so that either the microphone or the red disable light on the microphone button can have power, but not both at the same time. This ensures that you always have a visual indicator that the microphone is disabled, and that the indicator is reliable—i.e., if the light is red, the microphone is disabled. (By the way, if an Echo loses power while its microphone is disabled, when power is restored it will always start back up in the same condition.)

Remember that disabling the microphone only ensures Alexa can't listen *on that specific device.* Alexa will still be able to listen on any nearby devices unless you disable their microphones as well.

DISABLING THE CAMERA WITHOUT DISABLING THE MICROPHONE (ECHO SPOT OR ECHO SHOW)

Echo devices with a screen have both a microphone and a camera, and like the audio-only devices, they have a disable button, but this button disables both the microphone and the camera. If you want to disable one you may also want to disable the other, but there may be times when you want to just turn off the camera but leave the microphone working so you can continue talking to Alexa.

Fortunately, you can turn off the camera, and only the camera, in an Echo Spot or Echo Show via a voice command: *"Alexa, turn off the camera."* Note that when you use the disable button to turn off the microphone and camera together, you'll see a red line along the bottom of the screen (though the color arguably looks closer to orange) along with the "disable" icon at the lower right corner. But when you just turn off the camera, you don't get any visible indicator.

As mentioned in chapter one, the Echo Show 5 has a physical camera shutter which the larger Echo Show lacks.

WHISPER MODE

Whenever you speak to Alexa, it responds at whatever volume the Echo device was previously set to. But if you need to interact with Alexa late at night, early in the morning, or anytime there's someone sleeping nearby and didn't realize how high the Echo's volume was set, Alexa's response can be quite a jolt.

To avoid this unpleasant scenario, activate the Whisper Mode feature, allowing Alexa to whisper back at you when you whisper your request. You can turn this on from the Alexa app by tapping **Menu >**
Settings > Alexa Account > Alexa Voice
Responses, and then turning on **Whisper**

Figure 2-7: Whisper mode keeps Alexa's voice down when you whisper.

Mode. This is an account setting, so it applies to all your devices (Figure 2-7).

One caveat to be aware of is that using Whisper Mode requires a bit of finesse. You really need to whisper to Alexa (that hushed, breathy tone) to receive a response in kind. If you simply speak softly, Alexa will most likely not interpret that as a whisper and thus respond at normal volume.

CHANGE ALEXA'S VOICE ACCENT AND/OR LANGUAGE

When you use Alexa in the United States, it of course speaks in American-accented English by default. But if you'd rather Alexa speak to you in another language—or speak English but with a different accent—you can change Alexa's language settings to accomplish this on a particular Echo device. But first there are some important caveats to be aware of.

When you change Alexa's language, what you're usually changing is the country Alexa is configured to operate in, and this has ramifications. For starters, it means that you may lose access to certain Alexa features, because not every feature is available in every country. It also means some of your skills stop working, as Alexa skills must be specifically written to support specific languages. Lastly, some music or other audio content may no longer be available to play because it's not licensed for use outside the U.S.

Also keep in mind that if you're not fluent in the language you configure Alexa to speak (or your speech is heavily accented), it may have difficulty understanding you, and you may have trouble understanding Alexa as well.

With all those caveats out of the way, switching Alexa's language can still be useful if you or someone in your household would benefit from interacting with Alexa in a language other than English. It's easy to make the change from the

Alexa app—tap **Devices** > **Echo & Alexa**, the Echo device you want to change, then scroll down to **Language** (under **General**) and select the new language (Figure 2-8). Alexa has about a dozen country/language options including French, German, Italian, Japanese, and Spanish as well as various flavors of English (flavours?), such as Australia/New Zealand, Canada, India, and UK. (As of this writing, Alexa supported two versions of Spanish, Spain and Mexico, but Amazon has announced US Spanish will be an option sometime before the end of 2019.)

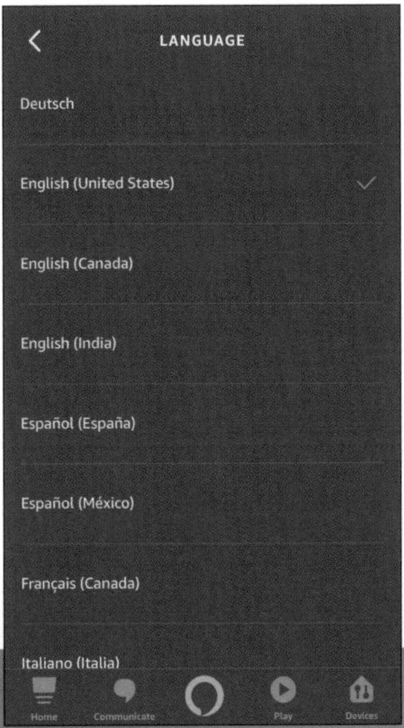

No matter what language you configure Alexa to speak, the voice is always female.

Figure 2-8: Change Alexa's voice accent and/or language (but beware the caveats).

USE VOICE PROFILES (SO ALEXA CAN TELL DIFFERENT PEOPLE APART)

Unless you live alone, chance are at least two people in a household will use Alexa. In most cases it's not necessary for Alexa to know who specifically is speaking in order to respond to requests, because after all, the answer to a question doesn't necessarily change based on who is asking it.

But there are some situations when knowing who is speaking will allow Alexa to provide customized responses when you're using some of the features we discuss later in this book. For example, if you use Alexa's Messaging feature to send a message to another Alexa user (more on this in chapter five) and Alexa recognizes your voice, Alexa won't need to ask who you are before playing your messages or sending messages on your behalf. Similarly, recognizing your voice allows Alexa to skip asking you for a security code each time you make purchases by voice (more on this in chapter seven).

To set up an Alexa voice profile for, say, a spouse or significant other, that person's account must first be added to the Amazon Household for the primary Alexa user. This lets the two accounts share things like Amazon Prime benefits and digital content. To see the status of your Amazon Household, visit www.amazon.com/myh/households, select **Manage Your Household**, and add the person's Amazon account, if necessary. (Note that Amazon allows a maximum of two adult accounts per household.)

Next, to have Alexa learn and recognize your voice, set up a Voice Profile from the Alexa app. Make sure you're signed into the Alexa app with your own Amazon account, then tap **Menu > Settings > Alexa Account > Recognized Voices**. Then tap **Your Voice**, the **Begin** button (Figure 2-9), and complete the steps shown, which will ask you to go to an Echo device (muting

Figure 2-9: Create a voice profile so Alexa can recognize you.

any other nearby devices first) and either repeat a series of phrases or confirm who Alexa thinks you are.

To verify that Alexa is identifying individuals correctly, say, *"Alexa, who am I?"* The response will include the name of who Alexa believes is speaking as well as which user's Amazon account Alexa is currently connected to. At any time if you find Alexa is connected to the wrong account, you can say, *"Alexa, switch accounts,"* and Alexa will connect to the other user's account.

A question we come across a lot is about how to configure Alexa to only respond to specific voices—that is, to respond only to "authorized" voices and ignore requests from all others including kids, houseguests, your grey parrot, etc. The short answer is that there's no way to do that today, though Amazon may choose to offer that capability in the future.

TALK TO ALEXA FROM YOUR COMPUTER

You already know you can talk to Alexa on an Amazon Echo or other Alexa-enabled device as well as through the Alexa app. But if you're not near an Alexa-enabled device and don't have your phone handy, you can also talk to Alexa from a (microphone-equipped) Windows or Mac computer:

Windows 10: Amazon actually makes an app for Windows 10, which you can find at www.microsoft.com/en-us/p/alexa/9n12z3cctcnz?activetab=pivot: overviewtab

MacOS: Use an app called Reverb for Amazon Alexa, which you'll find in the Mac App Store: itunes.apple.com/us/app/reverb-for-amazon-alexa/id1144697855?mt=12

You can even talk to Alexa from a computer without having to install software first—just point a web browser to Echosim.io. (Echosim.io is ostensibly a skill-testing tool for developers; if you're curious, flip the console switch at upper right of the page to ON to see the code that's generated with each Alexa interaction.)

3
KEEPING ORGANIZED

Alexa can be a great help keeping your daily life streamlined and well-organized. In this chapter we'll learn how Alexa can help you remember important information and events with features including lists, timers, reminders, and alarms.

LISTS

Those who strive to be as efficient as possible often do so with the help of lists. (My personal motto: "If it's not on my list, it doesn't exist.") Those with a similar philosophy will find that letting Alexa handle lists for you is a better alternative to scrounging for pen and paper or typing into your phone when you need to track things to do, items to buy, errands to run, or just about anything else around the house. (I often used an Alexa list as a quick way to record ideas that came to me while writing this book.)

Creating Lists

Alexa includes two built-in lists named "Shopping" (natch!) and "To-do," which may be the only lists you ever need Alexa to manage for you. But if not, you can create your own custom lists too.

To have Alexa create a list, say, *"Alexa, create a list."* Alexa will ask you what you want the list name to be, then repeat it back to you for confirmation. To save a step you can also have Alexa create and name the list with a single command: *"Alexa, create a {list name} list."*

Adding Items to Lists

After Alexa creates and names your list it will ask you what you want to add to the list. After you speak an item Alexa will repeat it back to you and ask whether there's anything else you want to add (the verbiage will vary slightly with each request to sound more, you know . . . human).

To add items to any list in the future, say, *"Alexa, add {item(s)} to my {list name} list."*

When adding multiple items to a list at the same time, it's a good idea to verbally separate each item with "and" so Alexa doesn't potentially misinterpret multiple items as a single item. This doesn't happen often—Alexa is smart enough to know, for example, that "ice cream" is most likely one item, not two. (Or maybe sometimes you actually do want to buy both ice and cream.) Although Alexa will try to use the context of words to determine whether they should go together or not, it doesn't always get it right, so you can avoid misunderstandings by saying "and" between individual items.

Reviewing, Editing, and Deleting Lists

To have Alexa tell you what's on a list from an Echo device, say, *"Alexa, what's on my {list name} list?"* Alexa will tell you how many items are on the list, tell you the five most recent items added, then offer the next five items, and so on.

On an Echo device with a screen, Alexa will respond similarly, and also show the list on screen—most recent item first—and read the entire list rather than five items at a time. If you'd rather look at the list than listen to it, tap the screen and Alexa will stop speaking but keep the list on screen. (You can also tap an individual item to edit it or swipe it to the left to mark it completed.)

If you want to see your list when there's no Echo device around (while at the store, for example), you'll find it in the Alexa app by tapping **Menu > Lists** (Figure 3-1). Tap a list to view it and correct any list mistakes, if Alexa heard something incorrectly, which you can't currently do with voice commands.

To have Alexa remove an item from a list, say, *"Alexa, check off {item} from {list name} list."* In lieu of "check off" you can also use the terms "delete" or "remove." Regardless of which term you use, this doesn't purge the item from the list, but instead marks it "complete." You can also mark all items on a list complete by saying, *"Alexa, clear my {list name} list."*

If you decide you don't need a given list anymore, you unfortunately can't have Alexa delete the list for you via a voice command. Instead, you must do it from the Alexa app, where it's a two-step process. Find the list you don't want anymore, swipe it to the left and tap **Archive**. To remove the list for good, tap on **View Archive** (Figure 3-2), swipe to the left on a list, and then tap **Delete**. Or, to purge all your archived lists, tap the three dots at upper right and then **Delete All**. (Note that once you delete an archived list it's gone for good, and that you can't delete the two built-in lists, Shopping and To-do.)

Figure 3-1: Manage your lists in the Alexa app.

Using Alexa with Third-Party List Apps

Alexa can communicate with several third-party list apps (including Any.do and Todoist), so if you already use one you may be able to link Alexa to it. From the Alexa app, tap **Menu > Settings > Lists** to see the third-party list services Alexa works with. Tap the one you want, and you'll be given the option to enable the

Figure 3-2: You must archive lists before you can delete them.

Alexa skill for that service as well as link Alexa to your account for it. Once Alexa is linked to the service, lists and items you add to lists via Alexa will automatically sync there, and any lists/items you create via that service's app will sync to Alexa's lists.

TIMERS

Once upon a time, our oven timer got a lot of use, and usually it had nothing to do with cooking or baking, but rather to track things like whether a child spent the requisite time on homework or reading, or count down how long before needing to leave the house to get to some appointment. Using a smartphone's timer app is a big improvement over the oven, but having Alexa create timers for you is a quantum leap over both.

Before we get into the details of how to use Alexa timers, it's important to remind you that they're device-specific, so the Echo device where you ask Alexa to create a timer is the only one that will sound off when the timer is finished. Similarly, you can only manage or get status on a timer from the device it was created on.

To have Alexa set a timer, say, *"Alexa, set a timer for x minutes (or hours)."* You can also name your timer: *"Alexa, set a pasta timer for 12 minutes."*

Naming your timers is obviously helpful when you need to keep track of multiple timers, or if you just have one timer when you want to avoid the possibility of forgetting why you set it when it eventually goes off (this has actually happened to me on a few occasions).

When you ask Alexa to set a timer, Alexa will repeat your timer duration and name, if you gave it one, and begin counting down immediately (not audibly, of course, because that would be super annoying).

Alexa can set timers up to 24 hours long (if you need one longer than that, use a reminder, which we'll cover a bit later in this chapter). You can specify the timeframe in minutes, hours, or, if necessary, both. Tell Alexa to set your timer for an hour-and-a-half or for 90 minutes, for example. Or for 11 minutes, or, if you *really* need precision, for something like 8 hours, 14 minutes, and 58 seconds.

Managing Timers

To check on your active timer(s), say, *"Alexa, timers."* Alexa will list all timers with names and time remaining on each.

To get the status of a specific named timer, say, *"Alexa, how much time on {timer name}?"*

When a timer goes off, an alert tone will play, and Alexa will tell you your timer is done. This will continue indefinitely about every three seconds until you tell Alexa to cancel or stop the timer.

To end a timer once it's gone off (or before), say, *"Alexa, cancel (or stop) timer."*

If you want to only stop a timer's countdown temporarily, you can tell Alexa to pause the timer, and to have Alexa start it again, tell Alexa to resume it. You can also ask Alexa to add or subtract time from a timer in progress, but you can't add to a timer once it's gone off.

Using Visual Timers

To get the status of your timers visually rather than by voice command, you have several options. The first is to create your times on an Echo device with a screen. When you do, the screen will show a timer countdown in the upper right corner until the timer goes off. When you have multiple timers running, only the one that expires first appears on the Home screen.

To have Alexa display all running timers on the screen, say, *"Alexa, show timers."* You can swipe left on a timer to delete it. Interestingly (annoyingly, actually) you can't pause a timer set from an Echo device with a screen—either from the screen or by voice command.

Remember the analog wall clocks—the ones with twelve numbers and two hands—we all used to have in our kitchens but that many of us stopped using years ago? That kind of clock may seem antediluvian in the era of voice-activated virtual assistants, but if you make heavy use of Alexa timers you might consider bringing Amazon's $30 Echo Wall Clock into your home.

This accessory pairs with an Echo device via Bluetooth and has 60 LEDs to provide a visual indication of how many minutes are remaining on a timer. For certain timer scenarios—in particular, I'm thinking of getting kids off to school each morning—it can be a lot more convenient than periodically asking Alexa for timer status updates (which, let's face it, most kids aren't going to do anyway).

Check Timers on the Alexa App

If you have timers set on multiple Echo devices, you can check their status via the Alexa app. Tap **Menu > Reminders and Alarms > Timers,** then select an Echo device at the top of the screen to see the timer(s) running on that device (Figure 3-3). Note that though you can pause or delete timers from the app, you can't create timers from the app, nor will it notify you when timers expire.

REMINDERS

While timers can only span 24 hours, when you need to remember things over a longer period, you can use Alexa reminders. (Like timers, reminders are device-specific, so the device where you create them is the device where Alexa will deliver the reminder.)

Figure 3-3: Check timers in the Alexa app.

To set a reminder, tell Alexa what you want to remember and when: *"Alexa, remind me to drop off dry cleaning Thursday at 10 a.m."* Alexa reminders must include a task name, day, and time, so if you leave any of these out, Alexa will ask you to provide them.

You can also have Alexa set recurring reminders: *"Alexa, remind me to take out the garbage every Tuesday at 8 p.m."* Recurring reminders can be daily, weekly, weekdays, or weekends, so, unfortunately, you can't set a reminder to pay a bill on, say, the 22nd or first Saturday of every month, or to work out on Mondays, Wednesdays, and Fridays.

To review your reminders, say, *"Alexa, what are my reminders?"* Alexa will respond with the next four upcoming reminders, then refer you to the app to review the complete list. (On an Echo device with a screen, Alexa will display a scrollable list of your reminders; as with timers, you can swipe left to delete reminders.)

You can also ask Alexa to tell you reminders for today, tomorrow, the weekend, or a specific day.

Reminder Notifications

When your reminder time arrives, Alexa will say, "I'm reminding you . . ." or "This is a reminder . . ." and cite whatever you asked her to remind you of. Note that Alexa will speak the reminder only twice and then fall silent (unlike timers, which continue until you tell Alexa to stop).

Remember that reminder notifications are delivered to the Echo device where they were created, so if you have multiple devices, it's a good idea to try to create a reminder on one where you're most likely to hear it when it's delivered (i.e., don't create it on the bedroom Echo device if you spend most of your time in the kitchen or living room).

But let's face it—unless you can predict the future, you don't necessarily know hours, days, or weeks in advance where you'll be when a reminder's time comes. Because you might be out of earshot of the Echo device when that happens, you can also receive your reminder on your mobile phone as a notification from the Alexa app—provided, of course, you have the app installed and that it's configured to send notifications.

Here's how to verify the Alexa app can send notifications on your mobile device: on Apple devices, tap **Settings** > **Amazon Alexa** > **Notifications**, then make sure **Allow Notifications** is turned on. On Android devices, tap **Settings** > **Apps** > **Amazon Alexa** and make sure **Notifications** is turned ON. (Note that the above steps are performed within settings of your mobile device, not within the Alexa app.) Then, open the Alexa app, tap **Menu** > **Settings** > **Notifications**, then and set **Reminders on this device** to **Enabled**.

Alexa can't speak your reminder out loud on a mobile device the way it does on an Echo device. Instead, Alexa notifications consist of an alert tone and a pop-up message that contains the actual reminder information.

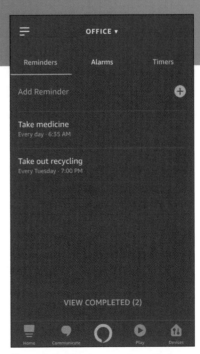

Use the Alexa app to edit the name, date, or time of any existing reminder as well as convert a standalone reminder to a recurring one (and vice-versa). You can also create new reminders from the Alexa app when you're not currently near an Echo device.

To review or edit reminders in the Alexa app, tap **Menu** > **Reminders & Alarms** > **Reminders**, and then at the top of the screen choose the Echo device the reminder was created on (Figure 3-4). You can also create reminders via the Alexa app and choose which Echo device it will be announced from, but you can't change it once the reminder has been created.

Figure 3-4: Check or create reminders in the Alexa app.

LOCATION-BASED REMINDERS

Creating reminders for a specific day and time usually makes sense for things like appointments and errands, but there are situations where it can fall short. Consider this scenario: You have a reminder set to do something for a time you normally get home from work, but you get stuck in heavy traffic or your plans otherwise change, and when Alexa's reminder sounds at the Echo device, you're not home to hear it.

Or course, your reminder will also appear as a notification on your mobile device, but what if the reminder is for a task you can't perform until you're home anyway? By the time you finally get home you'd likely forget about that notification, and thus the reminder, and thus the task, unless you somehow happen to remember to review the past notifications on your phone. (Remember that Alexa will only announce a reminder on an Echo device twice, not indefinitely as with timers.)

You avoid this scenario with Location-based reminders, which tie the reminder to your arrival or departure from somewhere instead of a fixed time. So how does Alexa know when you arrive or depart somewhere? Via the Alexa app, aided by the GPS location function of your mobile device.

Before you can use location-based reminders, you must confirm the Alexa app can access the location information from your mobile device: on Apple devices, tap **Settings** > **Amazon Alexa** > **Location** and make sure **Allow Location Access** is set to "Always." On Android devices, tap **Settings** > **Apps** > **Amazon Alexa** > **Permissions**, and make sure **Location** is turned on.

Defining Locations

The next step in using location-based reminders is to predefine locations of places you frequent (such as your work, school, etc.) and would likely want Alexa to use in a reminder. Although you can always specify your current location when creating a reminder, that obviously requires your presence at that location to create the reminder. Defining locations in advance gives you the flexibility to create reminders for locations before you expect to be there.

You can define locations in the Alexa app by tapping **Menu** > **Settings** > **Your Locations** (Figure 3-5). You'll already see an entry for your **Home** address (Alexa already knows this as part of your account information), as well as entry for **Work** that you can fill in, if desired. Create additional locations by tapping **Add Location** and typing the address or name of a place. If you enter a location's street address, you'll have to give the location a name before you can save it, but if you enter a place name, that will become the name of the location unless you edit it before saving (for example, changing something like "Smith County Middle School" to "Jimmy's School" or simply "School").

Creating Location-Based Reminders

Once you've defined your locations, you're ready to use location-based reminders, which you can do by specifying the location in lieu of a day and time. For example, say, *"Alexa, remind me to drop off my prescription when I get to the supermarket"* or *"Alexa, remind me to get gas when I leave home."*

Note that you can have Alexa create reminders upon arrival at a location even if you're currently there. So, for example, if you're already at home and say, *"Alexa, remind me to turn on the dishwasher when I get home,"* Alexa won't deliver that reminder until you leave home and come back there again.

Use the Alexa app to create a location-based reminder when you're away from home—tap **Menu** > **Reminders & Alarms** > **Reminders**, then **Add Reminder**, and remember to tap **At a location.** (Figure 3-6)

Because location-based reminders rely on features of your mobile device, it's important to be aware that many mobile devices automatically enter a "Do Not Disturb" mode when connected to a car's Bluetooth (or they detect a moving vehicle) which suppresses notifications for the

Figure 3-5: Defining locations allows you to use location-based reminders.

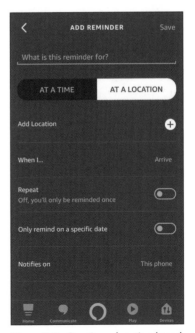

Figure 3-6: Creating a location-based reminder in the Alexa app.

purpose of eliminating driving distractions from your phone. The upshot is that if your mobile device is set to Do Not Disturb and you set a reminder to go off when you leave a location, the Alexa app will still issue the reminder notification, but you may not be aware of it until the next time you check your phone.

ALARMS

Compared to the usefulness of lists, timers, and reminders, alarms don't seem all that exciting. But most of us already use alarms to wake up every morning, and unlike on an alarm clock or even your phone, alarms on Alexa can be created and manipulated almost entirely by voice; no more fumbling for a button to stop or snooze an alarm, for example.

Alexa can set alarms (including recurring ones) for a specific day of the week, a given date, or weekdays or weekends only, and just like reminders and timers, alarms are device-specific.

To set an alarm with Alexa, you need only tell it the details. For example, you can say things like:

"Alexa, set an alarm for Friday at 6:15 a.m."
"Alexa, set a repeating alarm for every weekday at 7 a.m."
"Alexa, set an alarm for March 18th at 5:55 a.m."

To review your alarms, say, *"Alexa, check alarms"* and Alexa will list them. To delete an alarm, say, *"Alexa, cancel alarm"* (if multiple alarms exist Alexa will ask which one you want to cancel).

Snoozing Alarms

When an Alexa alarm goes off, it can be "snoozed" just like pretty much every alarm clock since time immemorial—say, *"Alexa, snooze"* to snooze an alarm for the customary nine minutes of extra shuteye. But if you want to snooze for less or more time—because after all, the nine-minute snooze is a relic from the 1950s when alarm clocks had mechanical gears—just add that to your voice command (e.g., *"Alexa, snooze for 30 minutes."*) By the way, "snoozability," for lack of a better word, is a major advantage of an Alexa alarm over most other types, including phones and phone apps.

For example, iPhone's built-in alarm lets you snooze for only nine minutes, and Android's makes you choose a snooze of 5, 10, 15, or 30 minutes. Various alarm apps give you more snooze options, but most still force you to choose from a fixed number of predefined intervals. But with Alexa you have ultimate snooze flexibility—you can snooze for five minutes, then four more, then twenty more. (Just promise you'll get up eventually, okay?)

Managing Alarms

Review and edit your alarms from the Alexa app by tapping **Menu > Reminder and Alarms > Alarms** and choosing a specific Echo device from the list at the top of the screen (Figure 3-7). You'll notice that any alarm you've created previously is listed here and if it was a one-time alarm, it's still listed, but just turned off. Tap an alarm to change the time, date, whether it repeats, or the sound that it plays. Sound choices here include a typical slate of sound effects plus numerous celebrity voice options (just in case you want to wake up to Alec Baldwin or Dan Marino for some reason).

When you're looking at an Echo device's list of alarms in the Alexa app, you can tap **Settings** at the bottom of the screen to configure various options (Figure 3-8) that will apply to all alarms on that device. These include the volume of the alarm alert (which is independent from what the Echo volume is set to) and whether you want the sound to gradually increase in volume.

Music Alarms

One last thing about alarms—if you'd rather wake up to music rather than a sound effect or celebrity voice, Alexa can oblige. Just ask

Figure 3-7: Add a new alarm or reactivate and edit an existing one from the Alexa app.

Figure 3-8: Configuring alarm volume.

it to set an alarm to a song, an artist, a musical genre, or a playlist. Say something like, *"Alexa, wake me up at 8 a.m. weekdays to Hall and Oates."*

Note that Alexa's ability to play specific songs will be dependent on its availability on whatever streaming services you've linked to Alexa. For more on this, see chapter six.

HAVE ALEXA REMEMBER ANYTHING FOR YOU

We've already seen how Alexa can help keep you organized and on schedule with Lists, Reminders, and Alarms. But when you just need to remember some random or impromptu bit of information, Alexa can help there too, saving you the trouble of having to find somewhere to jot it down (not to mention finding that information when you need it later).

To have Alexa remember something say, *"Alexa, remember {whatever you want to remember},"* and Alexa will commit it to memory. To recall what you asked Alexa to remember, you have a couple of options. One is to simply ask Alexa a question referencing the specific topic or fact you asked her to remember. For example, if you asked Alexa to remember, "I left my glasses on top of the washing machine," then you can ask, *"Alexa, where are my glasses?"* and Alexa will recite what you told it to remember about your glasses. You can also say, *"Alexa, what do you remember?"* and Alexa will reply with a rundown of every individual item you asked Alexa to remember.

Note that Alexa treats the things you ask it to remember as individual, unrelated items. It doesn't analyze the information you ask it to remember or compare new items with old ones, so two different "remembrances" can be contradictory. For example, if you tell Alexa to remember that your glasses are on the washing machine and then later tell it to remember your glasses are in the garage, Alexa doesn't update its memory of your glasses with the latest location. Instead, Alexa will remember both of those things and respond with both when asked about your glasses.

If you want Alexa to forget something you've asked it to remember, you

Figure 3-9: Ask Alexa what it remembers, then check your activity in the Alexa app to view the last and delete what you want it to forget.

can do that, but not through a voice command. (At least, not directly or entirely through a voice command.) To purge something you've asked Alexa to remember, first say, *"Alexa, what do you remember?"* then immediately open the Alexa app and choose **Menu** > **Activity** (Figure 3-9). The most recent entry will list all the stuff Alexa's remembering for you and let you delete (or even edit, via the pencil icon) each individual item.

4
CALENDAR AND EMAIL

If you're like most, you have an email account with a corresponding calendar that you use daily. If you link Alexa to your account for these services, it can check your messages and add appointments to your schedule via voice commands.

If you have personal email and calendar accounts from either Google (Gmail) or Microsoft, you can link them to Alexa. (Microsoft email accounts end in either @hotmail.com, @live.com, or @outlook.com.) Alexa also supports linking to Apple (i.e., iCloud) accounts, but only for calendar, not email.

Alexa can also link to business-focused Google and Microsoft email and calendar accounts (known as GSuite and Office 365, respectively).

PRIVACY AND SECURITY CONSIDERATIONS

Linking your email and/or calendar to Alexa can be very convenient, but because of the way Alexa works it's important to consider the security ramifications and decide whether the convenience is worth the potential risk to privacy. So, let's discuss those first before we go any further.

First and foremost be aware that by necessity, linking an email/calendar account to Alexa essentially gives Alexa unfettered access to the data in your account. If you carefully read the information that appears when Alexa requests permission to access your account (which we'll get to in a moment), you'll notice that it says that granting the permission will allow Alexa to "read, compose, send, and permanently delete all your email."

The second issue is who else besides you might be able to access your linked email or calendar account via Alexa voice commands. Although you can create a numeric code to control access to email, it provides limited security, and there's not even an option for a code with calendar access.

The bottom line is that if you're going to link Access to your email and calendar, you really need to trust the people you live with because there's no way to strictly limit access to the account holder alone.

LINKING AN EMAIL AND/OR CALENDAR ACCOUNT

To link an email and/or calendar account, open the Alexa app, tap **Menu > Settings > Calendar & Email** (Figure 4-1). Then tap **Add Account** and after choosing the type of account you have, follow the steps to link it to Alexa (which includes signing into the account and allowing permission for Alexa to access it). If you have multiple calendars on the account, you'll choose which specific calendars you want Alexa to access.

Note that for Google and Microsoft accounts, the default is to link both email and calendar on the account, but you have the option to choose only one or the other.

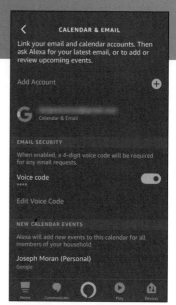

Alexa can link to multiple calendars on a single account and can review, move, or delete events on any linked calendar. But Alexa can only add new events to one calendar per account, so if you've linked multiple calendars, it's important to check which one Alexa will post new events to. You'll find this calendar listed under **New Calendar Events** (Figure 4-1).

Setting a Voice Code for Email Access

Once you've linked an email account to Alexa you should strongly consider setting up a voice code under **Email Security** (Figure 4-1). Alexa will require the 4-digit numeric code prior to

Figure 4-1: Link an account to allow Alexa to access your email and calendar.

executing any email-related commands, but fair warning—this code may not necessarily provide the security you might expect, because while the code is of course meant to be kept secret, it's a voice code after all. So, unless you live alone, there's a good chance that sooner or later you'll have to speak the code within earshot of someone. So much for secrecy.

That caveat notwithstanding, it's still worth setting up the voice code because some security is always better than none.

CHECKING EMAILS

Once you've linked an email account, Alexa can check it for new messages, or just check for messages from a particular sender. To check for new email, say, *"Alexa, check my email."*

When asking Alexa to check email, be sure to say "email" and not "messages," because the latter term refers to something different as far as Alexa is concerned; specifically, voice messages sent to you by other Alexa users. (For more on voice messages, see chapter five.)

Alexa will prompt you for the voice access code (if you've set one) then tell you how many new emails you've received within the past 24 hours. Then Alexa will report the sender and the subject line of the first message and give you the option to read, reply to, or delete the message, or say "next" to move on to the next message. If you don't say anything within five seconds, Alexa will automatically exit out of email. (Alexa only checks for emails received within the past 24 hours, so it won't tell you about any emails older than that—even if they're new and unread.)

If you're expecting an email from a specific person, you can save the time it takes to peruse all your new messages by asking Alexa to check for messages only from that person—say, *"Alexa, do I have email from {sender name}?"* Remember, Alexa will only check for new emails going back 24 hours.

Replying to Emails

If you tell Alexa to reply to a message, Alexa will ask you to dictate your response, and when you're finished speaking, read back your reply and ask if you want to send it now. If you say "No," she'll give you another opportunity to dictate a response.

When dictating an email reply, it's a good idea to keep it short and simple for several reasons. First, Alexa doesn't understand free form speech quite as well as the voice commands it's programmed to understand, so the odds of Alexa misunderstanding one or more words go up the longer your reply. Also, if you pause your speech for more than a second, Alexa tends to cut you off and start reading the reply back to you before you're finished. And speaking of reading back the reply, Alexa recites it rather quickly and without inflection, which can make it a bit hard to follow.

One more thing—Alexa doesn't include punctuation in your reply unless you dictate it (e.g., saying the words "comma" or "period") so to it's a good idea to do that to minimize potential confusion by the recipient.

Speaking of potential confusion, you can have Alexa automatically append a "Sent by Alexa" signature to the emails it sends on your behalf, which lets recipients know any questionable content in the email is probably the result of a dictation error. To add this signature, from the Alexa app tap **Menu > Settings > Email and Calendar,** the email account name, then scroll to **Signature** under the heading **Email Settings**. (Figure 4-2)

Alexa can only reply to messages you've received (and then, only those received in past 24 hours). Alexa can't initiate new outgoing email messages.

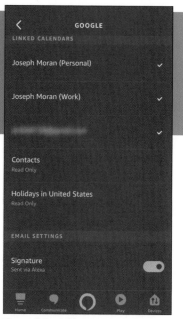

Setting Email Notifications

If you're waiting for an important email from a specific person, you can have Alexa automatically notify you when a new email arrives from that person, though you have to do it in a somewhat roundabout way.

First, ask Alexa if you have email from the person in question as described previously in "Checking Emails." Assuming you don't have any, Alexa will ask if you want it to tell you if that person sends you an email within the

Figure 4-2: Add a signature to your emails so a recipient knows it was dictated to Alexa.

next three days. If you say yes, when the next email from that person arrives, Alexa will light the ring on your Echo devices yellow indicating a notification waiting, and when you ask Alexa about it, you'll be told about the new email.

MANAGING CALENDARS

Once Alexa has access to your calendar, Alexa can tell you about your upcoming calendar events for whatever timeframe you choose. For example, you can ask, *"Alexa, what's on my calendar today?"* In lieu of "today," you can say "tomorrow," "this week," "next week," or a specific day or date (e.g., Saturday, or April 15th). You can also ask Alexa what's on your calendar for a specific day and time (e.g., *"Alexa, what's on my calendar Sunday at 3 p.m.?"*).

Alexa can not only tell you when there are events on your calendar, but also when there aren't events on your schedule, or when you have openings. For example, *"Alexa, am I available between noon and 2 p.m. tomorrow?"* or *"Alexa, when do I have an hour available on Thursday?"*

Adding, Rescheduling, and Deleting Calendar Events

To add events, you can say, *"Alexa, add an event to my calendar"* and have it prompt you for the specifics such as the day, time, and name of the event. Or, to save some time, include those particulars directly in your command—for example, *"Alexa, add Dentist to my calendar on April 21st at noon."* Either way, Alexa will ask you to confirm what you said before creating the event.

Remember that whenever you ask Alexa to create calendar events, they are added to the calendar specified under *New Calendar Events,* as described earlier. (Figure 4-1)

Although Alexa can only create new events on one calendar, Alexa can delete and reschedule events on any of your linked calendars—you don't have to specify the name of the calendar to do so; just tell Alexa the name, date, or time of the event.

To delete a calendar event, say, *"Alexa, delete my {event name, or date/time} event."* Alexa will repeat your request and ask for confirmation before deleting.

To move a calendar event, say *"Alexa, move {event name} to {date and/or time}."* If you specify a new day for an event but not a new time, Alexa will keep the original time.

5
COMMUNICATION

Not only can you talk *to* Alexa, you can also talk to other human beings *through* Alexa. In addition to functioning as a speakerphone of sorts, Alexa can serve as a messaging system between Alexa users, and if you have multiple Echo devices in your home, such as a room-to-room intercom or a whole-home public address system.

GIVING ALEXA ACCESS TO YOUR CONTACTS

A prerequisite to using most of Alexa's communication features is to give her permission to access the contact list from your mobile phone via the Alexa app. Here's how to do that depending on whether you're using an Apple or Android device:

- Apple: Tap **Settings** > **Amazon Alexa**, then turn on the **Contacts** switch
- Android: Tap **Settings** > **Apps** > **Amazon Alexa** > **Permissions**, then turn on the **Contacts** switch

When you give Alexa access to your contacts, the Alexa app will pull information from any contact lists that are set up on your phone. So, if you have multiple accounts set up—say, separate accounts for personal and work—Alexa will collect names and phone numbers from all of them. The Alexa app also periodically synchronizes your contacts, so as you add, delete, or modify them at their original source, Alexa's contact list will update accordingly. You can view all the contacts the Alexa app found by tapping **Menu** > **Contacts** (Figure 5-1).

It's important to note that Alexa syncs contacts one-way only, from your mobile phone to Alexa. This means you can't change a contact's information or delete a synced contact from within the Alexa app; you must do it from the contacts app on your phone. You do, on the other hand, have the option to create new contacts directly from the Alexa app (tap **Menu > Contacts**, the three dots at upper right, and then **Add a Contact**) and you're free to modify and delete these contacts as you need to. But always keep in mind the one-way sync—contacts you create in the Alexa app will only exist in the Alexa app—which means they don't sync back to your mobile phone contacts.

By the way, you don't have to create new contacts for your friends and family members that own Echo devices—Alexa already knows which of your contacts are also Alexa users. You may not necessarily know off the top of your head, but if you open up a contact and see "Alexa Calling & Messaging" as an option (Figure 5-2), that person is an Alexa user.

Nicknaming contacts

One thing you *can* do with an Alexa contact is assign it a unique nickname, which you can then use to identify the contact in voice commands in lieu of the contact's full name. It's a good idea to set nicknames for the people you expect to contact the most via Alexa, as it can save time and make it easier for Alexa to know who you're trying to reach. For example, nicknames can let Alexa know that when you say "Jack," you mean your friend John Smith, and "Frank" means your brother Frank, not Frank the landscaping guy.

Figure 5-1: The Alexa app can sync contacts from your mobile device.

Figure 5-2: Alexa Calling & Messaging under a contact name means that person is an Alexa user.

To assign a nickname to a contact, find the contact in the Alexa app and tap **Add Nickname** (Figure 5-2).

CALLING CONTACTS

To call a contact through Alexa you need to specify two basic pieces of information—who to call and how to reach them. And since you're doing this with your voice and not your fingers, your verbal instructions should be as specific as possible: *"Alexa, call {contact name} at {location}."* In this case, "location" can mean a specific saved phone number for your contact (e.g., home, work, mobile), or it can also be "Echo," meaning you want to call another Alexa user on his or her Echo device.

If you're calling another Alexa user and you both have Echo devices with screens, the call will include video as well as audio.

Note that you don't have to use prepositions and you can use possessives, so either way Alexa will know what to do if you say something like "Call John Smith home" or "Call John Smith's Echo." If there's ever any ambiguity about your calling request, Alexa will ask you confirm she's placing the call you intended. For example, if you just tell Alexa to "Call John" and you have multiple contacts named John, or one contact named John with multiple contact numbers, Alexa will ask you follow up questions to confirm the intended contact and how you want to contact them. But to save time and minimize possible confusion, when asking Alexa to make a call you'll generally want to state both your contact's first and last name (or nickname) and explicitly state the contact method as well.

LEAVING VOICE MESSAGES FOR ALEXA USERS

If you ask Alexa to call a contact's phone number and the person doesn't pick up, you'll naturally be sent to that phone's voicemail system to leave a message. Similarly, when you ask Alexa to call an Echo device, if the other person doesn't answer after about 30 seconds, Alexa will report the contact was not available and give you the option to leave a message.

But unlike a phone, if you'd like to skip the call and just leave the message, you can do that with Alexa by saying, *"Alexa, send {contact name} a message."*

Alexa will respond with "What's the message?" Tell Alexa the message, and after you finish speaking, she'll say, "Got it. Should I send it?" Say yes and the message will be delivered to your recipient.

Listening to an Alexa Message

When another Alexa user has left you a message, your Echo device chimes once and the light ring pulses yellow to indicate that either a message or some other kind of notification is waiting. There's no specific indicator for messages, so when you see the yellow ring say, *"Alexa, what are my notifications?"* to hear the notifications, followed by *"Alexa, play my messages"* to listen to the messages.

CALLING PHONE NUMBERS

Alexa calls are not limited to other Alexa users or people and phone numbers in your contact list. You can also have Alexa call phone numbers by specifying the area code and number you want to call. For example: *"Alexa, call 2 1 2 5 5 5 1 2 1 2."*

There are some limitations to the types of phone numbers Alexa can call, however. Alexa can't call 911 for emergency services, other x11 numbers (such as 311 for municipal government or 511 for traffic info), or 900 or any other toll numbers. Alexa also can't dial phone numbers outside of North America (i.e., United States, Canada, and Mexico). Lastly, Alexa can't dial by letter, so if your podiatrist's office number is 555-FOOT, you'll need to tell Alexa to call 555-3668 instead. These calling limitations apply both to numbers dialed directly and those stored within contacts.

For a way to get around this limitation and call phones worldwide via Alexa, see "Calling with Skype" later in this chapter.

NOTE: When using Alexa to call a phone, Alexa displays your mobile number as the caller ID to the recipient. You can turn this off in the Alexa app—tap **Menu > Contacts > My Profile and Settings** (under your name) and flip the switch next to **Show Caller ID** (Figure 5-3). By the way, although Alexa displays your mobile phone number for caller ID, it's not actually using your mobile phone or your carrier's network to make phone calls, so they are free.

SENDING TEXT (SMS) MESSAGES (ANDROID ONLY)

If you have an Android phone, you can send an SMS (a.k.a. "text") message to any contact's phone number using only your voice. To use this feature, you must first enable it in the Alexa app—tap **Communicate** > **Contacts** > **My Profile** and **Settings** and turn on **SMS Messaging**.

To send a text, say, *"Alexa, text {contact name} or {area code + phone number}."*

Alexa then asks you to state your message, and when you confirm you want to send it, your recipient gets the message both transcribed into text and with a link to the recorded audio.

Note that Alexa can't send group texts, texts to emergency services, or texts containing pictures or video. Also, unlike voice calling which doesn't use your mobile carri-

Figure 5-3: Alexa uses your mobile number for Caller ID unless you turn off Show Caller ID.

er's network, using Alexa to send texts does use this network, which is something to keep in mind if your carrier charges you for texts.

ANSWERING AND ENDING ALEXA CALLS

We've covered how to use Alexa to call out from your Echo device, but what happens when someone tries to call into your Echo device? When you receive an incoming call, your Echo device will chime, and the light ring will pulse bright green. Alexa will say "{Contact name} would like to talk," and you can then tell Alexa to either pick up or not by saying *"Alexa, answer"* or *"Alexa, ignore."* If you don't tell Alexa what to do it will automatically drop the call after about 30 seconds.

If you do choose to answer a call, to end it just say, *"Alexa, hang up (or end call)."* It's worth reminding you that unlike on a phone, with Alexa there's no button to press to end a call so it's important to remember to tell Alexa to hang up when a call is finished. Otherwise, you may remain connected without realizing it.

Remember: your cue that an Echo device is on an active call—and someone is listening on the other end—is the green spinning ring.

USE ALEXA DROP IN AS A TWO-WAY INTERCOM

Alexa's "Drop In" feature is a quick way to establish communication between a pair of Echo devices, essentially turning them into a two-way intercom. So, if you have multiple Echo devices at home, you can use them to Drop In from one room to the another. But where Drop In really gets interesting, is that with the proper permission, you can ask Alexa to Drop In on another Echo device that belongs to someone else—such as a family member or friend. This can be a very handy way to help keep tabs on an elderly parent, for example, even if they don't live with you.

The important thing to remember is that unlike a call you place through Alexa, a Drop In automatically connects you to another Echo device *without someone having to "pick up" on the other end*. For this reason, you can't Drop In on an Echo device that doesn't belong to you unless the device owner has granted you permission to do so. (To help distinguish Drop In from ordinary Alexa calling, it may help to remember the former as the feature with the potential to eaves*drop*.)

Drop In on Another Echo Device

To Drop In on an Alexa device that belongs to you, say, *"Alexa, Drop In on {device name}."* When you initiate a Drop In, the receiving device will chime briefly and within a second or two establish a two-way audio connection. The light ring on both devices will spin green to indicate communication in progress.

It's much easier to know where you're Dropping In to when your Echo devices have descriptive names. See chapter two for more information on how to name Echo devices.

To Drop In on someone else via their Echo device, say, *"Alexa, Drop In on {contact name}."*

If your contact hasn't granted you permission to Drop In on them, Alexa will tell you that and suggest you try calling instead.

Giving Permission for Someone to Drop In on You

If you want to give permission for someone to Drop In on you, open the Alexa app, tap **Communicate > Contacts**, choose the contact and turn on **Allow Drop In** (Figure 5-4). If you don't see the **Allow Drop In** option, it means the contact isn't an Alexa user.

This is a good time to mention that Drop In permission isn't reciprocal, so if Alice gives Bob permission to Drop In on her, it doesn't mean Alice can automatically Drop In on Bob. Bob still needs to give his explicit permission for Alice to Drop In on him.

Also, be aware that when someone Drops In on you, Alexa doesn't announce the name of the contact that's dropping in, and you won't be able to ask Alexa since it can't respond to questions while a Drop In (or a call) is in progress. Though you're always free to end a Drop In (or any other type of call) whether you initiated it or not, the takeaway is that for privacy reasons, you should be sure to only grant Drop In permission to people you trust not to abuse it, such as family and close friends.

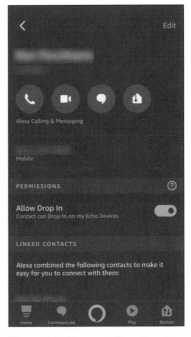

Figure 5-4: Turn on Allow Drop In to allow another Alexa user to Drop In on you.

USE ANNOUNCEMENTS AS A HOME PUBLIC ADDRESS SYSTEM

Alexa's Announcement feature turns your home's Echo devices into a public address system by giving you the ability to broadcast a message between one Echo device and all the others.

Let's say you have an important message to deliver to everyone in your home—something like, "dinner is ready," "it's time to leave," or "the cat is playing the piano again." Instead of screaming it at the top of your lungs or individually hunting down the members of your household, Alexa can record your message and broadcast it to anyone within earshot of an Echo device—say, *"Alexa, announce (or broadcast) {your message}."*

Alexa will chime and then play the recording of your message on your other Echo devices.

USING COMMUNICATION FEATURES FROM THE ALEXA APP

So far in this chapter we've covered various ways Alexa allows you to communicate from an Echo device. But you should be aware you can also use most of Alexa's communication features directly from the Alexa app. While calling someone from the Alexa app on a phone isn't all that different from simply calling them from the phone's own calling app, the Alexa app can be a very handy way to send messages and do Drop Ins and Announcements when you're away from home and not near your Echo device.

You can do all these things by tapping the **Communicate** tile at the bottom of the Alexa app (Figure 5-5). Unlike an Echo device where your voice is the only way to initiate communication, doing it via the Alexa app doesn't necessarily require your voice. For example, when you initiate a Drop In from the Alexa app, you choose the recipient from a list of all your Echo devices and contacts that have agreed to accept Drop Ins from you.

Similarly, when sending an Announcement from the Alexa app, you can type it in lieu of speaking it, in which case Alexa will read the announcement in its own voice. And if you do choose to record an announcement in your own voice from the Alexa app, you can preview it before sending (which you can't do from an Echo device).

Figure 5-5: The Alexa app provides access to all of Alexa's communication features.

The Alexa app can send announcements but not receive them, because you probably wouldn't want it to suddenly start playing audio from your pocket or purse.

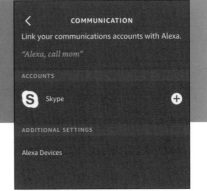

<COMMUNICATION
Link your communications accounts with Alexa.
"Alexa, call mom"
ACCOUNTS
S Skype   ⊕
ADDITIONAL SETTINGS
Alexa Devices

CALLING WITH SKYPE

A limitation of Alexa's calling features that we mentioned earlier in this chapter is that you can't use Alexa to call phones outside of North America. If you happen to

Figure 5-6: Link a Skype account to Alexa to call other Skype users (and phones outside of North America).

have a Skype account, you can link Alexa to it to circumvent that limitation. Once Skype is linked to Alexa, Alexa can use Skype to call any phone number worldwide (provided you have an active Skype subscription or calling credits available). And on an Echo device with a screen, you'll be able to place a video call to another Skype user.

To link Alexa to your Skype account, in the Alexa app tap **Menu** > **Settings** > **Communication** > **Skype** (Figure 5-6) and follow the steps to log into your Skype account and grant Alexa permission to access it.

To place a Skype call, say, *"Alexa, Skype {contact name}"* or *"Alexa, Skype {phone number}."*

Note that in this case, "contact name" doesn't refer to your mobile device contacts, but to contacts in your Skype account, which you can review and modify from the Skype app or website.

LIMITING OR DISABLING COMMUNICATIONS ON AN ECHO DEVICE

Just because you may want to use some of Alexa's communication features on your Echo devices doesn't necessarily mean that you want to use every feature on every device. Therefore, you can either temporarily or permanently restrict calls, Drop Ins, and Announcements on individual Echo devices as you see fit.

To access the Communications options for an Echo device, open the Alexa app, tap **Devices > Echo & Alexa**, tap the device you want, then choose **Communications** (Figure 5-7). From here you'll be able to turn off an Echo device's communications features entirely, turn off only announcements and/or Drop In capability, or simply limit who can Drop In on the Echo device. There are three Drop In options: **On**, the default, which means that contacts with permission can Drop In; **My Household**, which means only other Echo devices on your account can Drop In; or **Off**.

Figure 5-7: Enable or disable communication features on an Echo device.

USING DO NOT DISTURB

An alternative to disabling Alexa's calling features on an Echo device is to put it into "Do Not Disturb" mode, which will prevent it from receiving any incoming communication—calls, messages, Drop Ins, or Announcements.

To put an Echo device into Do Not Disturb mode, say, *"Alexa, Do Not Disturb"* and to take it out of Do Not Disturb, say, *"Alexa, end Do Not Disturb."* Note that the only visual indication that an Echo device is in Do Not Disturb is the brief purple flash of the light ring, and you only see that after Alexa responds to you. In other words, there's no obvious way to tell that an Echo device is in Do Not Disturb without talking to Alexa through it, though you can tell by viewing the device's settings in the Alexa app.

The Alexa app is also the way to toggle your various Echo devices in or out of Do Not Disturb without having to visit each

Figure 5-8: Schedule daily Do Not Disturb time for an Echo device.

one in person. To do so, tap **Devices** > **Echo & Alexa** > the device name, and scroll down to **Do Not Disturb**. From the app you can also schedule Do Not Disturb to automatically start and end at specific times each day, which you can't currently do via voice commands (Figure 5-8).

When an Echo device is in Do Not Disturb mode, it prevents the device from receiving calls, messages, Drop Ins, or announcements, but you can still use that device to initiate any of these things. Also, even if a device is in Do Not Disturb, it will still receive alerts for reminders, timers, and alarms.

6
ENTERTAINMENT

A big attraction of Alexa is as a source of entertainment, particularly music. In this chapter we'll outline various ways to use Alexa as a music system around the house.

LINKING A MUSIC SERVICE

If you're an Amazon Prime member—and if you use Alexa you almost certainly are—you already have access to Amazon's Prime Music service, which lets Alexa stream from a catalog of two million songs. That seems like a lot, and it is, at least until the first time your attempt to stream a song is unsuccessful because that song isn't included in Prime Music.

If you subscribe to Amazon's paid Music Unlimited service (which currently bills itself as offering "tens of millions of songs" for a cost of $10 per month or $79 per year) Alexa links to it automatically. But if you subscribe to any third-party music services such as Apple Music, Pandora, or Spotify, you can link Alexa to them as well and greatly expand the music library Alexa can draw from.

To link a music service from the Alexa app, tap **Menu > Settings > Music > Link**

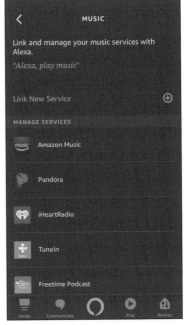

Figure 6-1: Alexa can connect to numerous non-Amazon music services.

New Service (Figure 6-1). Next choose a service and follow the steps to enable the Alexa skill for it, log into the service, and grant Alexa access to the account.

When you have multiple music services linked to Alexa, you can choose which of them will be the default music service. From the screen shown in Figure 6-1, scroll down to **Default Services** under **Account Settings** (Figure 6-2). (The default service is the one Alexa tries to play your music request from if you don't explicitly say which service to use.)

In addition to the names of songs, artists, albums, or playlists, you can ask Alexa to play music based on genre or characteristic (e.g., "disco," "Italian songs," or "up-tempo music").

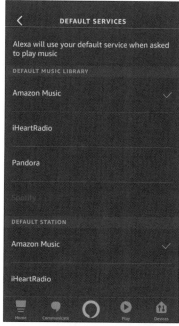

Figure 6-2: Choose which music service you want Alexa to automatically play from.

HAVE ALEXA ANNOUNCE EACH SONG

If Alexa is playing an unfamiliar song, you can ask it to name the song while it's playing but that requires Alexa to reduce the music volume temporarily in order to respond. Another option is to say, *"Alexa, turn on Song ID."* With Song ID enabled, Alexa will automatically report the song title and artist before each song plays.

Song ID works when you're playing from Amazon Music, but not from other music sources. And unfortunately, even if you're using Amazon Music, Song ID currently doesn't work when playing to a stereo pair or multiple rooms.

CREATE A STEREO PAIR FOR BETTER MUSIC QUALITY

While the audio quality of Amazon's Echo devices is quite good and gets better with each new generation (the current generation of devices are a big improvement over the previous ones), the one thing you won't get out of any Echo device is stereo sound because they all contain only a single internal speaker (plus, Echo devices aren't large enough to space speakers far enough apart for true stereo separation).

But while you can't get stereo sound out of a single Echo device, you can get it out of two of them if you configure them as a Stereo Pair. From the Alexa app, tap **Devices > Add (+) > Stereo Pair/Subwoofer**, and follow the steps to select two Echo devices (Figure 6-3) you want to link and specify which will be left and right channel. After a few minutes of setup time, your speaker set will be created, and when you ask Alexa to play music through either Echo device the output will be in stereo across both devices. (For best sound you'll want to have the two devices at least a few feet apart.)

Figure 6-3: Configure two identical Echo devices as a stereo pair.

Looking for songs to show off the stereo effect? Try Pink Floyd's "Wish You Were Here" or Queen's "Bohemian Rhapsody."

Note that you can only create a Stereo Pair from identical Echo devices, so you can't pair, say, an Echo and an Echo Dot or a 2nd and 3rd generation Echo Dot. Also, although you can create more than one Stereo Pair, you can't give Stereo Pairs customized names. Unfortunately, Alexa will automatically name the first Stereo Pair you create as, well, "Stereo Pair," the second will be "Stereo Pair 2," and so forth, which can get confusing, particularly if you want to create stereo pairs in different rooms.

While music is playing on a speaker set, any volume changes will apply to both Echo devices, but you may find it better to control the volume on Stereo Pairs via the Alexa app rather than by voice command, because doing the latter disrupts the music by temporarily reducing the song's volume on one Echo device (so Alexa can hear your command).

When not playing music, Echo devices that are part of a speaker set will still work autonomously, so when you speak to Alexa either one might respond depending on which one is closest to you.

You can use the same procedure for creating a Stereo Pair to link an Echo Sub to one or two Echo devices for enhanced bass. The $130 Echo Sub is not an Alexa-powered underwater vessel (though we're sure something like that is probably in the works), but rather, a 100-watt subwoofer accessory you can wirelessly link with an Echo, Echo Dot, or Echo Plus.

MULTI-ROOM MUSIC

It wasn't very long ago that the only way to have a whole-home music system was with bulky, power-hungry, and expensive audio components, wall- or ceiling-mounted speakers, and lots of in-wall cabling to connect them. More recently, wireless speaker systems from companies such as Sonos eliminated the wiring and the bulky components, but at $200 and up per speaker, not necessarily the high cost.

But Alexa's multi-room music feature lets you create an extensive and flexible whole-home audio using relatively inexpensive Echo devices. The best part of multi-Room music is that it allows you to mix and match different Echo devices, so you can use pricier Echo models in rooms where better audio quality is more important to you, and less expensive models—such as the extremely inexpensive Echo Dot—in areas where simply having a speaker is more important than how good it makes music sound. (Though for the record, the audio quality of the current 3rd-generation Echo Dot should be more than adequate for most non-audiophiles.)

For more on the speaker differences of Echo devices, see chapter one.

To use Alexa's Multi-Room Music, you must first organize your Echo devices into speaker groups. You can have multiple speaker groups and an Echo device can be a member of multiple groups, so you can start out with one group that includes all your Echo devices, add another that includes the ones in bedrooms, another with only upstairs Echo devices, and so forth.

To create a speaker group in the Alexa app, tap **Devices** > **"+"** > **Add Multi-Room Music Speakers,** then follow the steps to name the speaker group and choose which Echo devices will be part of it (Figure 6-4). It will take a few minutes for Alexa to create the group. After it's complete, repeat the process to create as many additional groups as you want.

To play audio on a speaker group, say, *"Alexa, play {your music choice} on {group name}."* To change volume in a speaker group, you have a couple of options. To set the volume for a specific device, speak to Alexa on the device whose volume you

Figure 6-4: Group Echo devices from different rooms together for multi-room music.

want to change. But to change volume for all devices in the group, reference the group name in your voice command. If you're playing on the Everywhere group, for example, say, *"Alexa, volume 4 everywhere."* Use the Alexa app to adjust the volume for individual devices without having to be near them—tap **Play**, then tap the speaker icon on the bottom banner that shows what's currently playing to reveal individual volume slider controls for each device.

When Alexa is playing multi-room music and you tell it to stop playing, it stops playing on all devices, regardless of which device Alexa heard you from.

Tap **Devices** in the Alexa app and scroll to the bottom to see all your multi-room music groups and/or stereo pairs listed under **Speaker Groups** (Figure 6-5). Tap a group name to edit it and change its devices.

ALEXA CAST

As convenient as it is to have Alexa play music for you by voice command, there will likely be times that you want to browse music before deciding what to play. It's a bit like going to a favorite restaurant—you may often know exactly what you want to order, but sometimes you just want to check out the menu first.

Perusing an exhaustive music catalog by Alexa voice command obviously isn't practical, but with Alexa Cast, you get the best of both worlds—browse and select music on your mobile device, and then send, or cast it to an Echo device to play. Alexa Cast works with either Amazon's Prime Music or Music Unlimited service, but currently not with third-party music services that Alexa links to.

Figure 6-5: Speaker groups listed under Devices in the Alexa app.

Alexa Cast requires the Amazon Music app for either iOS or Android, and it must be signed into the same Amazon account that your Alexa devices are registered to.

To use Alexa Cast, open the Alexa Music app then swipe up on the **Now Playing** icon at the bottom center of the screen to open it. Next, tap on the "Cast" icon at bottom right (Figure 6-6). This will display a list of devices which may include individual Echo devices, Stereo Pairs, multi-room music groups, and third-party Alexa-enabled devices. Select the item you want to connect to, and whatever you subsequently play from the Amazon Music app will play from the device or group you chose.

There are a few things to be aware of when using Alexa Cast. First, once you begin music playback you can close the Amazon Music app or even turn your mobile device off and the music will keep playing because the music is streaming to the Echo devices from the Internet, not directly from whatever device is running the Amazon Music app. That said, if you do keep the app open you can use it as remote control to pause or resume playback, advance to the next song in an album or playlist, or even change the playback volume. (You can also do all of these things via Alexa voice commands on the playback device.)

The Amazon Music app will stay connected to Alexa Cast until you disconnect it, so remember to do so if you plan to use the app to play music via headphones, etc. later. Otherwise, you might find yourself wondering why you're not hearing music, or someone else might be wondering why they are. If the previously mentioned "**Cast**" icon is blue, it means the app is using Alexa Cast; to disconnect, tap the "**Cast**" icon then the **Disconnect** button.

Figure 6-6: The Amazon Music app casting to an Alexa-enabled device. Note the blue "Cast" icon at the lower right.

CONNECT AN ECHO DEVICE TO A BLUETOOTH SPEAKER

Although every Echo device (except for the Echo Input) has an internal speaker, you can also connect an Echo device to an external speaker via Bluetooth. Although it may seem odd to pair something with a speaker to another speaker, it can make sense in certain situations.

One example is if you want to listen to music somewhere outdoors where plugging an Echo into an electrical outlet may be impractical, or the Echo's own speaker isn't loud enough to be heard over the noise of a party. Or perhaps you're poolside and don't want to risk getting an Echo device wet (or worse, have it fall into the pool). In these scenarios, it's worthwhile having Alexa route audio to an external speaker that's more powerful, battery powered, and perhaps even waterproof.

To pair an Echo device with a Bluetooth speaker, first put the speaker into pairing mode—check the user guide of the speaker for how to do that, and make sure the speaker is at least three feet away from the Echo. Then from the

Alexa app, tap **Devices > Echo & Alexa**, and the device you want. Next, tap **Bluetooth devices > Pair a New Device** (Figure 6-7). Wait for your device name to appear and follow the steps to pair it to your Echo.

Once you've successfully paired a Bluetooth speaker with an Echo, you can unpair and re-pair it with voice commands: say, *"Alexa, connect my speaker"* or *"Alexa, disconnect my speaker."*

When using a Bluetooth speaker with an Echo, keep in mind that Bluetooth's maximum range is generally about ten meters (33 feet) and that's assuming line-of-sight between the two devices. Obstacles like walls and windows (particularly laminated ones) will reduce this, so if using a Bluetooth speaker outdoors, you might need to temporarily relocate the Echo device closer to a window for best results.

Figure 6-7: Use Bluetooth to pair an Echo device to an external speaker.

PLAY YOUR PERSONAL MUSIC COLLECTION VIA ALEXA

Most of us can subscribe to a paid music service boasting a library of tens of millions of songs and call it a day, secure in the knowledge that it provides access to all the music we could possibly want. But what if you want Alexa to play a personal music collection amassed over time from sources such as rare CDs; bootleg or live material; or even music created by you, friends, or family members?

Until recently, Amazon allowed customers to upload their personal music collections to Amazon Music in the cloud for playback on PCs, phones, and via Alexa on Echo devices. Although Amazon has since discontinued that feature, there are still ways to have Alexa play your personal music collection, though they require a bit of extra effort and expense.

One option is to subscribe to a music service that both works with Alexa

Figure 6-8: My Media for Alexa's dashboard controls.

and lets you upload your own music to it. As of this writing, two services meet both of these requirements—Apple Music and Deezer—and each cost about $10 per month. But if you'd prefer not to subscribe to a music streaming service just to stream your own music, there's another alternative—run your own media streaming software at home. Two good options are Plex, which is free, and My Media for Alexa (Figure 6-8), which starts at $5 per year after a free 7-day trial.

Both Plex and My Media for Alexa work basically the same way. First, you download the media server software to a PC or Mac computer (other install options are available depending on the version, including Linux PC, Network Attached Storage (NAS) device, or Raspberry Pi) and configure the software to read from the folder(s) where your personal music is stored. Then, enable the media server software's Alexa skill and link it to your Amazon account.

We've used both Plex and My Media for Alexa successfully, though the latter is a little easier to set up for the less technically inclined. Downloads for My Media for Alexa are available at mymediaalexa.com/#section-3 along with helpful videos outlining the installation process. If you want to use Plex, you'll find the download at www.plex.tv/media-server-down-loads/ and important information about setting it up at support.plex.tv/articles/115000320808-getting-started-with-alexa-voice-control/.

Regardless of which media server software you choose, you will likely need to reference the appropriate skill in your voice command when you want to play from your personal music collection, i.e., *"Alexa, ask {Plex/MyMedia} to play {desired music}."*

LISTEN TO REAL RADIO STATIONS

As great as it is to have access to on-demand streaming audio with the precise songs, genres, or other content you want, sometimes you may just want to listen to the radio—as in actual broadcast radio stations (not to be confused with curated streaming music service playlists that are sometimes referred to as "radio"). But if you're like many people these days, you might not even have a working radio in the house. But with Alexa you don't need one because it can play live stations from your local area, around the country, or around the world.

Alexa automatically provides access to two services—iHeart Radio and TuneIn—which stream live radio stations along with many other kinds of audio content. To ask Alexa to play a radio station say, *"Alexa, play {station name or call letters}."* Call letters are the best way to identify a station because like the frequency, they're unique to the station. If the radio station you asked for is among the thousands available from TuneIn or IHeartRadio, Alexa will automatically play it.

But there are many stations that aren't available on either of these two streaming services. You can expand Alexa's access to available radio stations with a third-party skill from Radio.com. Enable the Radio.com skill and Alexa will check that service along with iHeartRadio and TuneIn when you ask it to play a station.

If you subscribe to SirusXM satellite radio, Alexa can play those stations if you link it to the service. (See "Linking a Music Service" earlier in this chapter.) Also, the optional TuneIn Live service ($3.99 per month or $2.99 for Amazon Prime members) provides access to a wide variety of sports broadcasts. For more info on TuneIn Live, see cms.tunein.com/listen/live/.

HAVE ALEXA SET AN AUTOMATIC SLEEP TIMER

If you have Alexa play something before you go to bed, you can ask for a sleep timer to stop playback after a certain amount of time. To set a sleep timer, say, *"Alexa, sleep timer in x minutes/hours."* Note that you must ask Alexa from a device that's currently playing audio—if Alexa hears you through a nearby device, it will set up a conventional timer (as described in chapter three) on that device named "sleep."

7

INFORMATION AND SHOPPING

Alexa can also come in handy when you need information about the weather, your commute, whether the Mets won last night (probably not), or the contact info for a local store. And if the store you want to shop from is Amazon, Alexa can even help you place orders.

GET DETAILED LOCAL WEATHER INFO

When you ask Alexa about the weather, the information Alexa provides is a bit limited, consisting of the current temperature, conditions, and forecast for your city, followed by the day's expected high and low temperatures. The weather info is also somewhat generic, applying to your city or town, but not necessarily your specific area, and it typically doesn't get updated quite as quickly as weather conditions change.

To significantly expand Alexa's access to useful weather info, use a third-party skill called Big Sky, which provides detailed weather data based on the latitude and longitude of your actual street address. You can enable this skill with the command *"Alexa, Enable Big Sky"* and you'll be referred to the Alexa app in order to configure permissions for the skill, such as permission to access to your street address and send you notifications.

Once you've configured the skill, to use Big Sky, say, *"Alexa, Big Sky."* Alexa's response will include the weather info mentioned previously but also provide the "feels like" temperature and how long before any precipitation is expected to begin (and/or how long it will last). Alexa will also ask if you want to hear more details, and if you say yes, it will respond with the current humidity, dew point, and wind information, as well as when the day's high and low temperatures occurred, followed by sunset and sunrise times.

If you don't need the full current weather rundown, you can narrow the scope of your question to focus on the specific information you want. For example: *"Alexa, ask Big Sky for the weather on Friday,"* or *"Alexa, ask Big Sky if it will rain tomorrow morning between 6 and 8 a.m."*

Big Sky offers a paid premium subscription (a free 7-day trial, then 99 cents per month or 80 cents per month for Amazon Prime members) which provides additional features including five customizable addresses and the ability to view animated weather radar on Echo devices with a screen. Say, *"Alexa, ask Big Sky about premium"* and it will walk you through the signup process.

CHECK TRAFFIC ON YOUR COMMUTE

It wasn't too long ago that checking traffic conditions for your commute meant seeking out a broadcast traffic report or opening an app to see road conditions on your route. Traffic apps have made that information more detailed and easier to access, but Alexa makes it even easier.

If you let Alexa know your commute destination(s), you can ask about your commute conditions before you get in the car. From the Alexa app, tap **Menu > Settings > Traffic** (Figure 7-1), where you should already see your home address listed as the **From:** location (though you

Figure 7-1: Give Alexa your locations to get traffic information on the route.

can change it if needed). Enter your commute destination in the **To:** location, and if you have an intermediate stop like dropping off the kids on the way to work, enter that address under **Add Stop**. (You can only add one stop.)

Once you've saved your address information, say, *"Alexa, how's my commute?"* Alexa will respond with an assessment of your commute conditions along with an estimated time and a preferred route.

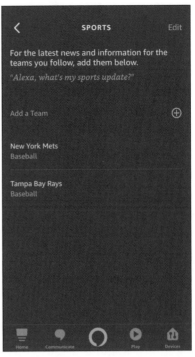

FOLLOW YOUR FAVORITE SPORTS TEAM(S) AND NEWS SOURCES

If you're a sports fan, Alexa's Sports Update lets you easily keep current on your favorite team, even if that means multiple teams across multiple sports.

Figure 7-2: Tell Alexa who you root for and it'll keep you up to date on your teams.

To configure Sports Update from the Alexa app, tap **Menu** > **Settings** > **Sports**. You'll already see one or more teams Alexa already chose for you based on where you live, so tap **Edit** if you need to remove any or all of them, then tap **Add a Team** to type and search for the ones you want to follow. (Figure 7-2)

Once you've specified your favorite teams, say, *"Alexa, what's my sports update?"* Alexa will report the most recent score, the day and time of the next game, and the latest news for your specified team(s).

Keep in mind that Alexa can give you scores and upcoming game info for any team, even if you haven't specified it as a favorite. Just say something like: *"Alexa, what's the latest {name of team} score?"* or *"Alexa, when's the next {name of team} game?"*

Alexa offers a similar update for news info called a "Flash Briefing." To configure it from the Alexa app, tap **Menu** > **Settings** > **Flash Briefing** > **Add Content** (Figure 7-3) to choose your news sources.

FIND AND CALL TELEPHONE NUMBERS OF LOCAL BUSINESSES

As easy as it is to find information about local businesses online these days, sometimes you just need to call the place. And if you've ever needed to get in touch with a local store or restaurant, the first step is typically going online to look up the phone number to call. But Alexa can save you some time and effort by not only finding the number for you but calling it too.

Say something like *"Alexa, find the number for Costco,"* and Alexa will do exactly that, responding with the phone number of the nearest business by that name as well as the address, so you know which location it found. Alexa will then ask if you want to call, and if you respond "yes" it will automatically place the call for you.

Figure 7-3: Alexa's Flash Briefing gives you a news digest from your choice of sources.

You can also modify your voice command to steer Alexa toward specific businesses that aren't necessarily the closest locations to you. For example, say, *"Alexa, find the number for Costco in {city name}."*

ALEXA VOICE SHOPPING

If you're an Alexa user, it's probably a safe bet that you regularly buy tangible stuff from Amazon too. Given that Amazon is a retailer as well as a technology company (and it was arguably the former before it became the latter) it should come as no surprise that Alexa lets you make purchases from Amazon through voice commands.

To be honest, we're a bit wary of using Alexa for online shopping because we don't really think shopping should ever become *too* easy. And the truth is that shopping by voice with Alexa isn't necessarily faster or easier than doing it the old-fashioned way—by visiting Amazon's website or firing up its mobile app.

But that said, there are situations when making purchases through Alexa can come in handy. Before we get into the details there are a few general things to be aware of when shopping through Alexa:

- It's only available to Amazon Prime members
- You can only buy Prime-eligible products that are sold or fulfilled by Amazon (the latter term means Amazon ships the item out of its own warehouse on behalf of another seller)
- Voice purchases are billed to your default Amazon payment method
- Orders of physical products are eligible for free returns (i.e., no deduction for return shipping)

Enabling (or Disabling) Alexa Voice Purchases

To shop through Alexa, you first need voice purchasing enabled on your Amazon account. Amazon typically does this automatically, but to confirm this (or if you'd rather turn voice purchasing off), go to the Alexa app and tap **Settings** > **Alexa Account** > **Voice Purchasing** and check the **Purchase by voice** setting (Figure 7-4). You may also want to tap **View 1-Click preferences** to see which credit card your purchases are billed to.

If you disable Alexa Voice Purchases, you can still use Alexa to search for products to buy or even add items to your Amazon shopping cart; you just won't be able to use Alexa to make the actual purchase.

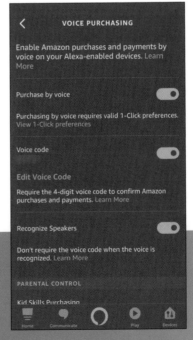

Figure 7-4: Configure Alexa's voice purchasing settings in the Alexa app.

Prevent Unauthorized Purchases with a Voice Code

If you choose to have Voice Purchasing enabled, you probably don't want it to be available to literally anyone in your house with a voice—which it will be unless you take precautions. Prevent unauthorized purchases with a four-digit code that will be required to confirm any purchase requests. Set up the code from the Alexa app by choosing **Menu > Settings > Alexa Account > Voice Purchasing > Voice Code** (Figure 7-4).

Here's the catch, though—not surprisingly, you must speak your voice code aloud, which makes it a challenge to keep it secret if you're not alone when you make a purchase. But fortunately, there's a solution to that dilemma—once you choose your four-digit code, the Alexa app will present an additional option called **Recognize Speakers** (Figure 7-4). When you turn this option on, Alexa will ask you for the Voice Code the first time you make a purchase, but it won't be required for subsequent purchases if Alexa recognizes your voice. (For Alexa to recognize your voice you must have a Voice Profile set up. For more on this, see chapter two.)

Making Voice Purchases

There are a lot of variables at play when making purchases through Alexa, and far too many potential scenarios to effectively cover here. But here are some general guidelines to keep in mind when shopping by voice.

When you say, *"Alexa, buy {item/product}"* Alexa's response and action will be different depending how general or specific your request is, as well as whether you've purchased that item or product before. If you have purchased that exact item before, Alexa will automatically add it into your Amazon online shopping cart. If you haven't, but the purchase request is specific enough, Alexa will also put the item in your cart. In other cases, Alexa will put a reminder in your cart in lieu of a specific item; check your cart from the Amazon app or website, and you'll be able to browse items that Alexa thinks meet the description of what you asked for. Alexa will let you add up to twelve of the same item to your cart.

To review what's in your cart at any time, say, *"Alexa, what's in my cart?"* To place an order for an item in your cart, say, *"Alexa, check out {item name}."* Note that although you can buy multiples of the same item as described above, you can only order one item at a time through an Alexa voice command. In other words, if you add three boxes of crayons to your cart, you can buy them all at once, but if you add a box of crayons and a coloring book, you must check each item out separately.

After you place an order, Alexa will confirm it and tell you the amount charged, the expected delivery date, and give you an opportunity to cancel the order, but this only works immediately after you place the order.

> In many cases, rather than doing your Amazon shopping entirely via Alexa, you'll probably find it more convenient to just have Alexa add items and reminders to your shopping cart, then head to the Amazon app or website to review and place the order.

Receive Amazon Shipping and Delivery Notifications

If your household is anything like mine there's usually a shipment from Amazon (and maybe several) en route at any given time. Regardless of whether you made the Amazon purchases through Alexa or the old-fashioned way, Alexa can tell you the status of those shipments, saving you the trouble of having to check for shipping confirmation emails or log onto the site.

Say, *"Alexa, where's my stuff?"* and Alexa will report shipments on the way to you along with expected dates of arrival. By default, Alexa will only refer to "shipments" without naming the purchased items; this is to avoid the possibility of accidentally divulging a gift or other surprise purchase. To make Alexa tell you what's actually in the boxes headed your way you need to change a setting in the Alexa app— choose **Menu > Settings > Notifications > Amazon Shopping**, then under **Give Ordered Items' Titles**, turn on one or both of the options: **Within Where's my stuff requests** and/or **Within Delivery notifications** (Figure 7-5).

Figure 7-5: Alexa can let you know when Amazon boxes are coming or have already arrived.

As that latter option implies, Alexa can also help keep you a step ahead of so-called porch pirates (those terrible people who swipe packages from doorsteps) by notifying you when a package from Amazon is out for delivery that day and/or has been delivered. Turn either or both options on under **Delivery Notifications**.

When a package is out for delivery or has been delivered, your Echo device(s) will flash yellow to indicate a shipping notification is available.

8

ALEXA AND
THE SMART HOME

Once you're using Alexa for things like organization, communication, entertainment, and so on, the next logical step is to use it as the cornerstone of a smart home setup for control of lights and other connected devices.

As of mid-2019, Amazon reports there are more than 60,000 smart home devices, from 7,400 different brands, that can be controlled via Alexa (and hence are considered "Alexa-certified"). This includes light bulbs, wall switches and outlets, door locks and doorbells, thermostats, cameras, and many other kinds of devices. For some perspective as to how fast this number of Alexa-compatible devices is growing, consider that at the beginning of 2019 those numbers were 28,000 devices from 4,500 brands.

For a current list of all Alexa-Certified products, see developer.amazon.com /alexa/connected-devices/compatible

While you don't need Alexa to have a smart home—only a connected device and the manufacturer's app by which to control it—Alexa can make a smart home, well, smarter by giving you centralized and hands-free control of smart devices compared to whipping out an app (or more likely, one of several apps—one for each different device manufacturer) each time you want to do something.

In this chapter, we're going to cover the essential elements of using Alexa to control smart home devices, which starts with knowing what kind of devices to buy, and by extension which kind of Echo device can control them. Then we'll move on to the nuts and bolts of naming devices, creating device groups, and more.

UNDERSTANDING SMART DEVICE HUBS

Smart home devices, by definition, need to connect to the Internet, and they generally do this in one of several ways. The first is to connect directly to a Wi-Fi network, the same way your Alexa-enabled devices and most other Internet-connected devices ranging from televisions to tablets do. If not Wi-Fi, smart home devices generally get their connections via either Zigbee or Z-Wave, which are wireless mesh networks designed specifically for smart home devices (because they don't use a lot of power and don't need to transmit a lot of data). Unlike Wi-Fi smart home devices, those based on Zigbee or Z-Wave need to communicate with a separate device called a hub or bridge, which it links to your Internet connection. Zigbee and Z-Wave are different technologies, and while they can coexist, they're incompatible with each other.

While a detailed comparison of Zigbee and Z-Wave is beyond the scope of this book, here's what matters as far as Alexa is concerned: as you may recall from chapter one, some Echo devices—namely the Echo Plus and the Echo Show (but NOT the Echo Show 5)—have a Zigbee Hub built in, which means Zigbee devices can connect to them without the need for a separate Zigbee hub.

In other words, if you don't have either an Echo Plus or an Echo Show, when shopping for smart home devices stick to those labeled "Works with Amazon Alexa" (which means they use Wi-Fi and will work with any Alexa-enabled device). If you do have one of the two previously mentioned Echo devices, you can expand your smart home device options to include Zigbee products, since you already have the required hub built into your Echo device. But you won't be able to control Z-Wave devices with Alexa—or in fact control them at all—unless you also purchase a Z-Wave hub. Some smart device hubs, such as the Samsung SmartThings Hub, are compatible with both Zigbee and Z-Wave, which give you the flexibility to use devices of either type.

To review, smart home devices that use Wi-Fi can talk to Alexa directly. Devices that use Zigbee or Z-Wave talk to a hub, which in turn talks to Alexa.

And in the case of the Echo Plus and Echo Show, that Zigbee hub is built-in, so you don't need a separate hub.

USING THE AMAZON SMART PLUG

Before we get into the further into details of how to manage smart home devices via Alexa, a few words about a specific device you're likely to find very handy. It's the $25 Amazon Smart Plug (Figure 8-1), which works directly with Alexa (i.e., no hub required) and is an easy and inexpensive way to make a "dumb" device smart.

To learn more about and/or purchase an Amazon smart plug, see www .amazon.com/Amazon-Smart-Plug-works-Alexa/dp/B01MZEEFNX /ref=sr_1_1?keywords=alexa+smart+plug&qid=1560803623&s=gat eway&sr=8-1

The concept behind the Amazon Smart Plug is simple—plug one into a wall outlet, then plug your device (such as a floor or table lamp, a fan, etc.) into the smart plug, and thus turn the device on and off by controlling power to the plug—no smart light bulbs or switches required. While the Amazon Smart Plug is particularly useful for lights, you can use it to control other kinds of electrical devices too. As you start building your Alexa-based smart home, you'll probably find use for at least one Amazon Smart Plug, if not several. (By the way, plenty of other companies offer Alexa-compatible smart plugs on Amazon, many of which cost less than the Amazon Smart Plug.)

Keep in mind that any smart plug will only work with devices that can be turned on and remain in the "on" condition when they lose power. A surefire

Figure 8-1: The Amazon Smart Plug
(image credit: Amazon)

way to test for this is to turn a device on, pull the plug for several seconds, then plug it back in again. If it's still on, you're good to go, but if you need to turn it back on again you won't be able to control it with a smart plug.

Another caveat to keep in mind when using a smart plug is that if you plug it into a power outlet that's controlled by a wall switch, that switch must be turned on and left on so that Alexa can turn power on and off at the smart plug. If the power is turned off at the wall switch, there's no way for Alexa to turn it on at the smart plug. So, if possible, it's best not to use smart plugs on switch-controlled outlets, and if it's not avoidable, make sure everyone knows the switch must be left on (or better yet, label the switch accordingly).

SMART LIGHTS—PLUGS, BULBS, AND SWITCHES

One of the first devices you'll likely want to control with Alexa are lights, and you can make your existing lights "smart" in one of three ways. If the light plugs into a wall outlet (e.g., a floor or table lamp), you can use a smart plug as just described. An alternative to a smart plug is to use a smart lightbulb, which you can also use in wall and ceiling fixtures that are controlled by a wall switch. Smart bulbs give you the additional benefit of letting you dim the light, and depending on the bulb, change its color. But when using either smart plugs or smart bulbs, the same caveat applies—when there's a wall switch involved, it must always remain on for Alexa to turn those devices on and off.

Using a smart wall switch is a way to avoid this problem, as it allows Alexa to turn power on and off directly at the switch. But the catch is that smart switches replace your existing switches and so need to be wired into your electrical system—if you go this route, for safety's sake we recommend you get an electrician to do the work rather than attempting it as a DIY job.

ADDING SMART DEVICES TO ALEXA

Regardless of whether you're dealing with a smart plug, bulb, switch, or any other smart device, the setup process through Alexa is basically the same. To add a smart device to Alexa, from the Alexa app tap **Devices** > "+", then choose the category and brand (not necessarily in that order) of the device you have and follow the steps provided (Figure 8-2). Depending on the type of device and its manufacturer, these steps may include downloading the device manufacturer's own app to perform initial setup and configuration, though in other cases you may be able to set up the device entirely from the Alexa app (particularly if

Amazon is the device manufacturer or if you are setting up Zigbee devices).

After the device is discovered, which can take up to 45 seconds, you'll have the option to add the device to a group—always a good idea, though you can skip it for now (read more about groups later in this chapter)—and if appropriate for the device, enable the device manufacturer's skill as well as provide account login information.

When you add devices to Alexa, the row of icons at the top of the Alexa app that appears when you tap **Devices** will expand to include whatever device categories you have, such as Lights, Plugs, Thermostats, etc. (Figure 8-3).

Figure 8-2: Adding a device to Alexa.

CONFIGURING DEVICE NAMES

When adding smart devices to your home, it's important that each device have a unique and descriptive name that references the device's location and purpose. Otherwise, you'll likely have a tough time giving Alexa clear instructions about what to turn on or off. Using smart light bulbs as an example, if you set up three bulbs—two in a hall ceiling fan and one in a den table lamp, the default

Figure 8-3: The Alexa app displays categories of the device types you've added.

names of those bulbs will be something like Light 1, Light 2, and Light 3—which is not too informative. In contrast, names like Hall Fan 1, Hall Fan 2, and Den Lamp clearly let you know what you're trying to control, so you can in turn tell Alexa. But don't get too descriptive—you may be saying these names a lot, so keep them as concise (and easy to pronounce) as possible.

If you've set up a device via a manufacturer's app, you probably had the opportunity to create a custom name. If not, once Alexa has discovered your device, you can rename it from the Alexa app. Tap **Devices**, then tap the device category (or swipe right for the **All Devices** category), then choose the device you want to rename. Tap the gear icon at upper right, then tap **Edit Name** to rename your device (Figure 8-4).

You can also ask Alexa to rename devices by voice: say, *"Alexa, rename {current device name} to {new device name}."*

ALL ABOUT SCENES

In addition to the above device categories, you may also see one labeled Scenes. Scenes are predefined device settings that you create in the device manufacturer's app (often a number of scenes are created automatically for you as well). An example of a scene for lights could be: turn on certain lights, then set them to blue and 40% brightness. For a universal remote control, a typical scene might involve turning on the TV and setting inputs to watch the cable box, and/or tuning it to a specific channel.

Figure 8-4: Tap Edit Name to change the name of a device.

When devices linked to Alexa include scenes, you'll find them listed under the Scenes category. You can't create scenes in the Alexa app—you can only do that from the device manufacturer's app—but you can add the scenes to Alexa groups and routines, both of which we talk about later in this chapter.

DEVICE GROUPS

So now that you've named each of your smart devices, are you going to remember all those names? Probably not, but don't worry because you don't necessarily have to. Even if you could, you won't always want to control individual lights, especially when they're all in the same room or part of a multiple-bulb fixture. Enter Device Groups, which organize multiple related devices so you can turn them on and off all at once, and in the process reduce the need for you to remember individual device names.

Here are some examples of how this works. Let's say you have three pendant lights suspended above your kitchen counter, each one with a smart bulb in it. You've named these bulbs Pendant 1, Pendant 2, and Pendant

3, and presumably want to always turn these three lights together (as you currently do from wall switch) rather than individually. By adding all three bulbs to a device group called "Pendants," you now have a way to control them together. In this example, you'd say, *"Alexa, turn on pendants."* (Of course, you could still ask Alexa to turn on the individual lights by name if you wanted to.)

Let's extend this example a bit further. In the previously mentioned kitchen, you also have four recessed ceiling lights in addition to the three pendants, and you'd like to be able to turn on both sets of lights at the same time. Adding all seven lights to a device group called "Kitchen" gives you the ability to do just that, by saying, *"Alexa, turn on Kitchen."*

Now let's extend this example even further. Back in our hypothetical kitchen you have an Echo device. Include this Echo device in your Kitchen device group, and when you're talking to Alexa through that specific Echo device, to turn on all the lights in that group you won't have to specify the name of any devices or the name of the group. You can simply say, *"Alexa, turn on lights,"* and Alexa will know which lights you're referring to by the Echo device you're speaking to. (To turn on the Kitchen lights from a different Echo device, you'd still refer to the group name.)

Generally, it will make sense to name and organize most of your device groups by the room your devices are in, but device groups don't necessarily have to be confined by four walls. For example, there might be a specific device (probably a light) you don't usually want to come on with all the others in a room. You might have several lights from different areas of your home you want to operate simultaneously. Or you might want to group together a handful of smart plugs that you use for outdoor holiday lights. There are a million different usage scenarios for device groups, but you get the idea.

That said, it's a good idea to organize your smart devices into groups even if you are starting out with so few devices that you don't have more than one device in any group. Why? By establishing your groups early on, you'll be less likely to need to reorganize or rename those groups as you add devices to them over time.

Creating Device Groups

To create a device group from the Alexa app, tap **Devices** > "**+**" > **Add Group**. You'll be asked to choose or create a name for the group, then from the list provided, specify which Alexa-enabled devices, smart home devices, or scenes you want to add to it (Figure 8-5).

Once you've created a group, you'll see it listed (alphabetically) in **Devices** under the **Groups** heading (Figure 8-6). Each group banner will include a light button and or a plug button depending on which types of devices are part of the group. Tap the button to turn the devices on/off in lieu of voice control or tap any other part of the banner to view and edit the group's device list and individually configure or turn off/on a device.

Note that smart devices such as lights can be members of more than one device group, but an Echo or other Alexa-enabled device can only be part of one group at a time. Currently there's no way to move a device from one group to another within the Alexa app other than to add the device to a new group and remove it from the old one. You can, however, move devices between groups via a voice command: say, *"Alexa, move {device name} to {group name}."*

Figure 8-5: You can add both smart home and Alexa-enabled devices to a group.

Figure 8-6: Device Groups you create are displayed under the Devices section of the Alexa app.

When you add a smart plug to Alexa—made by Amazon or anyone else—it can operate in one of two modes: plug (the default) or light. Choose the latter if the plug will be used with any kind of light—this will allow it to turn on with the other lights in the group when you say, *"Alexa, turn on lights."*

ROUTINES

Previously in this book we've discussed Follow-up Mode, which lets you give Alexa multiple questions or commands following a single use of the wake word. Routines take this concept a step further by having Alexa perform a series of actions automatically following a single command.

For example, consider a series of actions you might perform upon waking up in the morning—turning on lights, getting a weather report, checking traffic on your commute, and playing some music. These four actions would normally require four separate commands to Alexa but put those actions into a routine and Alexa will do it all in response to a single command or phrase, such as *"Alexa, Good Morning."* You can also configure the routine to occur automatically at a specified time without you having to say anything at all. In lieu of a specific time, you can also configure a routine to occur at indefinite or variable times such as sunrise or sunset, or when you arrive at or leave from a certain location. Routines can also be triggered when you dismiss an alarm, or even press a button (specifically an Echo button, which Amazon sells as a $20 accessory).

To set up a routine, tap **Menu > Routines > "+"**. Tap **When this happens** and choose what you want the routine to be triggered by (e.g., a custom phrase), then tap **Add action** and follow the steps to choose from a list of things Alexa can do in response to the trigger. In addition to the actions described just above, they can include things such as reading your calendar; saying something; controlling a device, group, or scene; and more. You can add as many actions to a routine as you want and if you don't want all your actions to immediately occur one after the other when the routine starts, use the **Wait** action to add one or more delays to the routine.

Before saving the routine and at any time afterward, you can edit the routine to reorder or delete actions, change the trigger, or disable the routine (Figure 8-7).

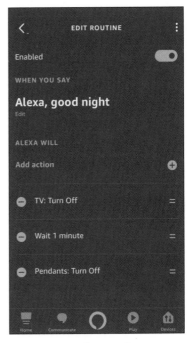

Figure 8-7: An Alexa routine to turn off the TV, and then after a short delay, the lights.

GUARD

As we discussed earlier in this book, Alexa must always be listening in order to respond to questions and commands. With a feature called Guard, you can take advantage of this to turn Alexa into a rudimentary security system by putting Alexa's vigilant ear to work when you leave the house.

With Guard, instead of simply listening for the wake word, Alexa will also listen out for and notify you upon hearing sounds that could signal potential trouble, such as breaking glass or smoke or carbon monoxide alarms. Alexa can also automatically turn your connected lights on and off, so it looks like someone's home even if the house is empty.

Before we go any further, it's important to note the things Guard can't do. It can't notify the police or fire department in the event of a problem, and it also can't work if your power and/or Internet connection aren't working. So, to be clear, Guard can't take the place of a conventional monitored alarm system (though it can forward notifications to supported security systems—as of this writing, SimpliSafe, Ring Alarm, and ADT Pulse are compatible with Guard).

> Guard works on Echo devices, but not third-party Alexa-enabled devices.

To set up Guard, open the Alexa app, tap **Menu > Settings > Guard > Set Up Guard**. Follow the steps to specify whether you want Alexa to monitor sounds from smoke or carbon monoxide alarms, glass breakage, and control of your lights. Since there's no point in manipulating lights during the day, if you choose to have Alexa control the lights, you'll be asked for your ZIP code so Alexa will know when sunset is in your area each day.

Guard only works when you tell Alexa you're away from home, which you can do by saying, *"Alexa, turn on Guard."* The light ring on your Echo device(s) will spin white to indicate that Alexa Guard is turned on. (Echo devices with a screen show a shield icon in the lower right corner.) Should Alexa hear something of concern while you're gone, you'll immediately receive a "Smart Alert" notification. Open the notification in the Alexa app and you'll be able to play back a ten second clip of what Alexa heard, and if you've allowed it, Drop In on the Echo that heard it to get a live audio and/or video feed from the device.

Note that to receive Guard Smart Alerts, be sure your Alexa app has permission to issue notifications on your mobile device. (See "The Alexa App" in chapter one for more info.)

To turn off Guard, say, *"Alexa, turn off Guard."* You can also activate and deactivate Guard by saying, *"Alexa, I'm leaving/I'm home,"* or from the Alexa app—the Devices section now includes a "Guard" banner and a button that toggles between Home and Away.

When you first set up Guard, it doesn't

Figure 8-8: Turn Guard on and off from this banner, as well as by voice command.

let you pick and choose which specific Echo devices or connected lights will be used by the feature—it automatically includes all your Echo devices as listeners and enlists all your configured lights. You can, however, customize Guard settings to take some of those devices out of the mix. Tap the banner (not the button) shown in Figure 8-8, then tap the gear icon at upper right.

From here (Figure 8-9) you can remove specific Echo devices or connected lights from Alexa Guard duty and turn off individual features such as alarm or glass break detection.

GET THE ROOM TEMPERATURE FROM AN ECHO PLUS

As we mentioned back in chapter one, one Echo device—the Echo Plus—has a built-in temperature sensor. Although it's intended to be used as a trigger for a routine, you can also use it to get your indoor temperature even if you don't have an Alexa-enabled thermostat.

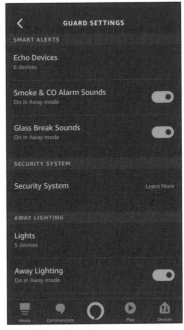

Unfortunately, you can't get the indoor temperature by asking Alexa, "What's the indoor temperature," and if you

Figure 8-9: Alexa Guard can monitor for troublesome sounds and/or control lights when you're away from home.

simply ask "What's the temperature," you'll get the outdoor temp. But here's a workaround—to get the indoor temperature, add the Echo Plus to a Device Group (as described earlier in this chapter), then say, *"Alexa, what's the temperature of {group name}?"* To keep the Echo Plus temperature sensor readings as accurate as possible, keep it away from things like air vents, space heaters, direct sunlight, or drafty windows.

LOCAL VOICE CONTROL

As we learned back in chapter two, Alexa's ability to understand and respond to your questions and commands doesn't reside within the device you're speaking to, but rather within the Alexa Voice Service (AVS) way across the Internet. In other words, when an Echo device can't connect to the Internet, it can't do anything for you other than tell you that it can't connect to the Internet.

Or can it? While the loss of an Internet connection turns most Echo devices into a paperweight, those which contain a smart home hub—namely, the Echo Plus and the Echo Show—include a feature called Local Voice Control which allows them to perform a handful of tasks without Internet access. This includes operating smart home devices (such as lights, plugs, and switches)

Figure 8-10: Turn on Local Voice Control to let hub-equipped Echo devices control lights and do some basic things without the Internet.

that are connected to its internal hub (i.e., Zigbee devices, but not Wi-Fi ones). It can also give you the time or date, as well as stop alarms, reminders, or timers that were set prior to the loss of Internet.

Local Voice Control can be enabled or disabled on devices, and although it has most likely been enabled by default on Echo devices that support it, you can confirm this by checking from the Alexa app—tap **Devices > Echo & Alexa**, the name of your device, then scroll down the list settings until you find **Local Voice Control**. (Figure 8-10)

Amazon is likely to expand the capabilities of Local Voice Control with future updates. If you ask a Local Voice Control–equipped Echo device, *"Alexa, what can you do offline?"* while it's offline, it'll respond with a current list of its capabilities. (Don't bother asking that question while the Echo is connected to the Internet; Alexa will just refer you to the Help section of the app.)

9
PRIVACY, SECURITY, AND PARENTAL CONTROL

In this chapter, we'll discuss what information Alexa collects and stores and what you can do to limit it. In today's world, technology use and privacy concerns often go hand-in-hand, so it's not uncommon to have questions about the privacy implications of using a voice assistant such as Alexa. Using Alexa can make your life more convenient, but the simple fact is that convenience and privacy occupy opposite ends of a spectrum—the more you have of one, the less you have of the other.

As much as we'd like to say that you can use Alexa without giving up any privacy, that's simply not the case. Let's state it plainly—the cost of Alexa's convenience is paid in privacy. Depending on which features you use, Alexa can know things like your interests (based on the questions you ask), your calendar and schedule, who you communicate with, when you come and go from your home, and that you love disco music.

Though there's no practical way to use Alexa without sharing this kind of information—potentially including some you may not necessarily intend to share—giving up some privacy isn't necessarily detrimental if you view it as a worthwhile trade for the varied capabilities Alexa gives you. The important thing is to be aware of what Alexa knows about you so you can exercise some control and make an informed decision about it.

UNDERSTANDING WHEN AND WHY ALEXA RECORDS

Alexa-enabled devices are commonly referred to as smart speakers, but they're more accurately called smart microphones. It's a common belief among some Alexa users that it doesn't start listening until hearing the wake word, but of course Alexa must first be listening in order to hear the wake word. So, in effect, Alexa is always listening, though it's not necessarily doing anything with what it hears.

As you may recall from chapter one, as soon as Alexa hears the wake word it begins recording audio and streams it to AVS for processing. In fact, according to Amazon, recording actually begins "a fraction of a second" prior to the wake word. The fact that Alexa is essentially a "hot mic" doesn't mean it records everything it hears, but there may be times Alexa records things such as random conversations and background sound.

How might this happen? One reason is because the wake word was used outside of the context of a request or command. For example, you might tell a family member, "I told Alexa to remind us when it's time to leave." Or you may use one of the alternate wake words—Amazon, Computer, or Echo—any of which are arguably more likely to come up in common parlance than "Alexa." Sometimes Alexa simply misunderstands speech or some bit of ambient noise as the wake word.

In these kinds of situations, Alexa may record some audio and send it out for processing before ultimately realizing that it didn't contain an actual request (and you might not be aware that it happened). Or Alexa may mistakenly hear a request when none was made and provide a response or take an action you didn't intend and you might not know what triggered it. Either way, Alexa will have recorded a snippet of audio that wasn't meant for it.

Although there's no practical way to prevent Alexa from mistakenly recording audio—other than not using "Alexa"—it is possible to review and delete what Alexa recorded.

Currently Alexa will only process what is spoken after the wake word, but Amazon has already filed a patent on a method to process speech that occurs *before* the wake word. Don't be surprised to see that capability added to Alexa in the future, most likely as an optional feature due to the privacy implications.

Review and Delete Voice Recordings

Everything Alexa records is saved, so you can go back and listen to what Alexa heard—or thought it heard. To review your Voice recordings, from the Alexa app tap **Menu** > **Settings** > **Alexa Privacy** > **Review Voice History** (Figure 9-1). Here you'll see a list of your voice interactions with Alexa starting with the most recent. The **Date Range** selector is set to **Today** by default, but you can set it to go back a day, a week, a month, to a custom timeframe, or your entire history of recordings.

For most entries, you'll see a transcription of what Alexa heard along with the time and device on which Alexa heard it. Others may be labeled "Unknown" or "Text not available – audio was not intended for Alexa." (The latter instances are those when Alexa thought someone was talking to it and then decided that wasn't the case after all.) You can tap the arrow to the right of any entry to expand it and play back the recorded audio or delete the recording.

If you want to delete voice recordings in bulk, you can do that too. When you select a **Date Range** as described above, a link immediately below will allow you to delete all the recordings for time period specified, though the app will warn you that doing so may "degrade your experience."

Note that when you have several Alexa-enabled devices in proximity, you may see a lot of entries that simply say "Alexa," which is the result of more than one device hearing the wake word.

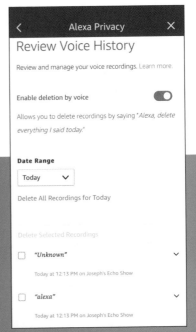

Note that you can also turn on an **Enable deletion by voice** option to delete the day's (and only that day's) recordings via a voice command. As you'll be warned when you activate it, however, this lets *anyone* delete what Alexa heard that day, which means it can potentially be used by someone to wipe out the record of their own Alexa activity on your account.

Figure 9-1: Review (and delete) your Alexa voice history.

Limit Amazon Employees' Access to Your Alexa Voice Recordings

If you've ever called the customer service line for a company you do business with, you probably remember hearing the disclaimer "Calls may be monitored for quality and training purposes" (or something like that). Well, it turns out Amazon does this sort of thing too.

In fact, there was recently a moderate uproar on the Internet following a news report that Amazon had "thousands" of employees and contractors worldwide who were tasked with listening to, transcribing, and annotating a sampling of customers' Alexa recordings to aid in the development of new features.

If you don't like the idea of your Alexa conversations potentially being heard by a human being working for Amazon, make

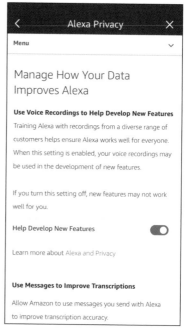

Figure 9-2: Amazon uses your Alexa recordings for product improvement unless you tell it otherwise.

a change to your privacy settings from the Alexa app. Tap **Menu > Settings > Alexa Privacy > Manage How Your Data Improves Alexa** (Figure 9-2). Then turn off the **Help Develop New Features** setting, and, optionally, **Use Messages to Improve Transcriptions** (which refers to messages you send to other Alexa users).

When you try to disable this setting, the app will warn you that new Alexa features may not work as well for you if you do.

HEAR AN AUDIBLE ALERT WHEN ALEXA PROCESSES A REQUEST

To minimize the possibility of Alexa interpreting random speech or noise as a request without your knowledge, you can configure an Echo device to play brief tones that alert you to when Alexa is recording and streaming audio for processing.

From the Alexa app, tap **Devices,** the name of the device, **Sounds,** then under **Request Sounds,** turn on either **Start of Request, End of Request,** or both (Figure 9-3). The **Start of Request** sound plays as soon as you say the wake word (or Alexa thinks it heard the wake word) so you know that audio is being recorded. The **End of Request** tone plays once recording has stopped.

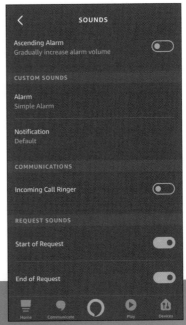

Start and *End of Request* tones must be configured for each device you want them on.

Figure 9-3: Request sounds let you know when Alexa is listening (and recording) on an Echo device.

VIEW AND MANAGE ALEXA SKILLS' ACCESS TO PERSONAL DATA

Over time you'll probably install numerous Alexa skills—maybe dozens of them—many of which will have permission to access some of your personal data. It's not always obvious what permissions a skill has—it's easy to gloss over when enabling a new skill from the Alexa app, and permissions don't even come up when you enable a skill via voice command. For this reason, it's a good idea to be aware of what permissions, if any, your Alexa skills have. Here are two ways to do it.

To see what permissions a specific skill has (not all skills have or need permissions), from the Alexa app tap **Menu** > **Skills & Games** > **Your Skills,** and then choose a skill. If you see a **Settings** button, tap it. If the skill has any permissions, you'll see them listed here, along with a **Manage Permissions** option at the bottom to enable/disable them (Figure 9-4). If you don't see a **Settings** button, the skill doesn't have any configurable permissions.

There's also a way to view and modify all the skills that have a specific permission (e.g. the ability to know your name or your street address). From the Alexa app, tap **Menu > Settings > Alexa Privacy > Manage Skill Permissions** (Figure 9-5). Tap on a permission to see which, if any, of your skills have that permission, and if desired, use the slider button next to the skill name to revoke its permission.

SECURELY DISPOSING OF AN ECHO DEVICE

Before you sell, trade-in, give away, or recycle an Echo device, it's important to deregister or reset it to return it to factory condition so a new user can't potentially use it to access your Alexa account.

To deregister an Echo device while it's still up and running, open the Alexa app, tap **Devices > Echo & Alexa**, the name of the device you want, then scroll down to **Registered To** and tap **Deregister** (Figure 9-6).

If you don't have the Alexa app handy, you can also reset the Echo device directly:

- On the 3rd-generation Echo Dot, press and hold the action button for 25 seconds.
- On the 2nd-generation Echo and Echo Plus, press and hold Microphone Off and Volume Down for about 20 seconds.

When the deregistration/reset process is complete, the Echo will power back up, ready to be configured as if for the first time.

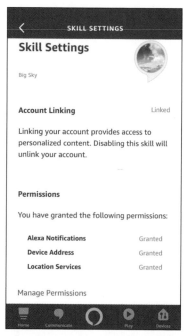

Figure 9-4: Permissions of an Alexa skill.

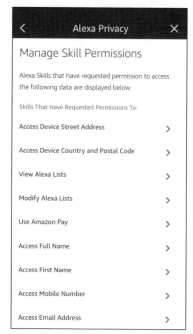

Figure 9-5: See which Alexa skills have a specific permission.

Different Echo devices can have different reset procedures, so for details on how to reset a specific model, see https://www.amazon.com/gp/help/customer/display.html?nodeId=GK7P5SPCQ3MN65VR

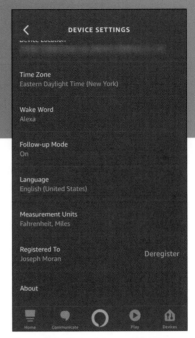

Figure 9-6: Deregister an Echo device to remove it from your account.

ENABLE TWO-STEP VERIFICATION ON YOUR AMAZON ACCOUNT

Throughout this book we've demonstrated that the Alexa app not only gives you direct voice access to Alexa, it's also Command Central when it comes to configuring what your various Alexa-enabled (or connected) devices can and can't do. What's stopping someone from logging into their own Alexa app with your Amazon account information to get access to your devices?

Perhaps less than you think. For starters, your username is hardly a secret since it's your email address, and if like many folks (far too many), you've chosen an easy-to-guess password (say, your child's name and year of birth), or worse, your Amazon password is the same one you use at other Internet sites (and thus, many places where it might fall into the wrong hands), there's not much of a barrier to prevent someone from gaining access to your account for mischievous or malicious purposes.

The best way to protect yourself against this is by enabling Two-Step Verification (TSV) on your Amazon account, which ensures that your password alone won't be enough to successfully log into your account. With TSV turned on, access to your account will require your password *and* a unique (and temporary) numeric code, sent to your phone or generated through an authenticator app.

Having to enter a code after your password can admittedly be inconvenient—extra security usually is—but it's a very effective way to protect your account from unauthorized access. Plus, the first time you enter a TSV code on a given device, you have the option to "trust" it so you won't have to enter codes again on that device.

To turn on Two-Step Verification, go to www.amazon.com/a/settings /approval and log into your Amazon account to access **Advanced Security Settings.** Click the **Get Started** button and follow the steps—you'll provide the phone number where you want to receive codes and decide whether you want to get them via text message or voice call. You also have the option to scan a QR code with a smartphone authenticator app from Google, Microsoft, or others and have the authenticator app generate your verification codes, which is even more secure than getting numbers via text or call.

Regardless of how you choose to receive your TSV codes, be sure to pay close attention to the information provided about how to log into Amazon from a device or app that doesn't directly support TSV codes (i.e., one that doesn't display a separate page after the password for you to enter the code). In these cases, your first login attempt will fail with an error message, but you will be able to log in successfully on the second attempt by appending your TSV code to the end of your password. (For more details on this, visit www.amazon .com/gp/help/customer/display.html?nodeId=201962400.)

ENABLING PARENTAL CONTROL ON AN ECHO DEVICE

Alexa can bring a seemingly endless array of information and other content into your home. That's great if you're an adult, but it can be a cause for concern when you have kids, particularly young ones. After all, you probably wouldn't want to give a young child access to a computer, smartphone, or tablet without some way to restrict what the youngster can see and do, and Alexa is no different. Fortunately, Amazon's FreeTime feature gives you a way to control how—and when—a child can interact with Alexa on a specific Echo device.

Enabling Amazon FreeTime (which is free) lets you do things like filter explicit lyrics in music, set daily time limits for Alexa use, and disable smart home control on an Echo device. It also automatically prevents purchases of products or content by voice. Amazon also offers Prime Members a paid version of FreeTime called FreeTime Unlimited, which in addition to the restrictions outlined above, includes access to curated and age-appropriate content and kid-friendly skills at a cost of $3 per month for one child or $7 per month for up to four.

To set up FreeTime from the Alex app, tap **Devices > Echo & Alexa**, the name of the Echo device you want, tap **FreeTime**, and turn it on. From here you'll be asked to specify the name and age of your child(ren), log into your Amazon account (you may be asked to enter a security code), and acknowledge a parental consent agreement. Then you'll be able to configure basic restrictions for explicit music filtering and enable or disable Alexa's communications features such as calling or Drop In. (Although it's not explicitly noted, you can't create or update lists from an Echo device with FreeTime enabled.)

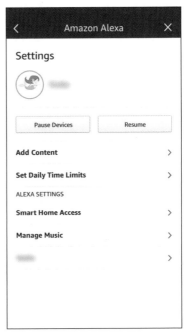

Then, tap the child's name under **Parent Dashboard Settings** to configure additional restrictions including daily time limits for using Alexa (Figure 9-7). Tapping the **Pause Devices** button will let you immediately lock down any FreeTime-enabled Echo devices for anywhere from 1–12 hours. When you pause devices, Alexa immediately says, "All done for now. We can play

Figure 9-7: Configure time limits and other parental controls for Alexa via FreeTime.

again later." and responds to any requests with "Sorry, I can't play right now. Try again later."

You can also configure Parent Dashboard settings from a web browser at parents.amazon.com.

Amazon offers an Echo Dot "Kids Edition" (regular price $70) which consists of a specially decorated (rainbow or blue) Echo Dot, a two-year warranty that covers physical damage, and a one-year subscription to FreeTime Unlimited. See more at www.amazon.com/s?k=echo+dot+kids+edition&crid=KI-ZLHN3CA109&sprefix=echo+dor+kids%2Caps%2C167&ref=nb_sb_ss_sc_1_13

10
EXTRAS

CREATE, SHARE, AND PUBLISH YOUR OWN ALEXA SKILLS WITH SKILL BLUEPRINTS

Authoring a conventional Alexa skill requires a fair amount of coding expertise as well as a degree of experience programming for a voice-centric interface. But Amazon has given those of us who lack that kind of knowledge a way to create simple yet useful Alexa skills via Skill Blueprints.

Skill Blueprints are basically templates—dozens of them organized into various categories such as *Learning and Knowledge*, *Fun and Games*, and *At Home*—which allow you to create customized skills for different scenarios by picking from menus and filling in the blanks to determine what Alexa will do and say. For example, there are Skill Blueprints for tracking workout routines, creating study aids such as flashcards and quizzes, creating a repository of information for babysitters or pet sitters, tracking chore responsibility among members of a household, and creating various games.

You can use Skill Blueprints to create as many skills as you want for personal use and share those skills with family or friends. If you think your skill may have broad appeal, you can even publish it to the Alexa Skills Store so that any other Alexa user can find and use it.

To get started with Alexa Skill Blueprints, visit blueprints.amazon.com (Figure 10-1) and browse the available blueprints. When you find one of interest, click on it, then click the **Make Your Own** button to kick off a series of steps to personalize questions Alexa will respond to, what Alexa's response to them will be, and, depending on which kind of skill blueprint you're working with, various other aspects of how Alexa will interact with the user of the skill. At the end of the process you'll choose a name for your skill (an invocation name, to be specific, which we talked about back in chapter one).

Don't worry about getting everything about a skill right the first time, because you can go back and edit or rename the

Figure 10-1: Build your own Alexa Skills using blueprints found at blueprints.amazon.com.

skill at any time. Note that you'll need to log into your Amazon account to create your new skill, at which time Amazon will classify your account as a "developer" account, which will allow you to share or publish your skill should you decide to. It can take several minutes (we've sometimes seen it take five or more) for your skill to be created and ready to use. Once it is, you can use it just like you would any other Alexa skill and you'll see the skill listed in the Alexa app under **Skills & Games > Your Skills**.

Once you've successfully created your skill, note the **Share with Others** and **Publish to Skills Store** options, but there are a couple of important things to note about sharing skills. First, in order to share a skill, you must stipulate that your skill complies with Amazon's content guidelines as well as specify whether the skill is intended for children under 13 years old, and if so, it will be classified as a kid's skill and require a parent's permission to enable. (The latter is a requirement of COPPA—the Children's Online Privacy and Protection Act.)

Second, if you initially use **Share with Others** to share your skill and later decide to publish your skill to the Alexa Skills Store, you'll need to first revoke share access to the skill. This means anyone you've shared the skill with won't be able to use it again unless they find and enable it from the skills store once it's published.

Third, publishing to the skills store requires some extra steps you don't have to go through when casually sharing a skill. You'll need to categorize it, select keywords so people can find it, create both brief and detailed descriptions, and possibly tweak or change the skill name you chose.

Once you've done all that and submitted your skill for publication, the review process starts, and it can take a couple of days. You'll hear back from Amazon via email regarding whether your skill was approved. If not, the notification will tell you why the skill was rejected and what you should do to fix it before submitting your skill again.

EXTEND ALEXA'S REACH WITH IFTTT

Usually when you want Alexa to perform some task, you'll find a feature, setting, or skill that lets you accomplish it in a relatively straightforward way. But sometimes that's not possible. When you come across something Alexa can't do natively or via a skill there may still be a way to accomplish it by extending Alexa's capabilities using a service called IFTTT.

IFTTT (which stands for If This Then That) is a free web-based service that uses "applets" to enable communication between different devices, services, and apps that can't directly connect to each other. Applets consist of one or more triggers with corresponding actions and are essentially conditional statements that say: "If x happens, do y" (hence the name of the service). By setting up an IFTTT account and linking it to your Alexa account, IFTTT can act as an intermediary—you say something to Alexa (the trigger), Alexa communicates with IFTTT, and IFTTT communicates with the device or service to perform the desired action.

Here's an example of how this can work—let's say you want to text someone your Alexa Shopping List. Although you can do that manually from the Alexa app, Alexa can't do that for you automatically via a voice command. But put an IFTTT applet into the mix, and suddenly asking Alexa what's on your shopping list can automatically trigger a text message to the recipient of your choice.

Sign up for IFTTT by visiting ifttt. com/amazon_alexa. You can create an account directly with IFTTT, or sign in using your existing Google or Facebook account. (You can also download an IFTTT app for your smartphone/tablet.) There's an endless roster of Alexa-compatible IFTTT applets available (Figure 10-2), but if you

Figure 10-2: An IFTTT applet to text an Alexa shopping list.

can't find one that does what you want, you can try to create your own (choose **My Applets**, then **New Applet** or "**+**").

Here are some examples of IFTTT applets you might find useful:

Have Alexa text your shopping list

ifttt.com/applets/cBkh79yu-ask-alexa-what-s-on-your-shopping-list-and-she-ll-text-it-to-you

Sync your Alexa To-Do list with your Google Calendar

ifttt.com/applets/UpRgMNhm-automatically-sync-your-amazon-alexa-to-do-list-with-your-google-calendar

Email your Shopping List to one or more Gmail accounts

ifttt.com/applets/284243p-tell-alexa-to-email-you-your-shopping-list

Receive a phone call when an Alexa alarm goes off

ifttt.com/applets/Y3fWvBGw-call-your-phone-when-your-amazon-alexa-alarm-goes-off

TAKE AN ECHO DEVICE ALONG WHEN YOU TRAVEL

Once you get accustomed to all Alexa can do, you may find it hard to go without it when you're away from home for an extended period for business travel, a vacation, or even just a long weekend visiting family. Although if you have the Alexa app you always have a version of Alexa with you, pulling your phone out and opening the app every time you want to talk to Alexa can be a pain, plus it's not a very practical way to share Alexa access with travel companions.

Bringing an Echo device along on the road is a much more convenient option, especially for things like streaming music, or to use certain features that aren't directly supported by the Alexa app (including reminders, alarms, and timers, all of which must be associated with an Echo device).

Configuring an Echo device to work outside your home requires changing the Wi-Fi network it connects to, and this in turn essentially involves repeating the process used to set up the Echo device on your home network for the first time.

To get an Echo device up and running on a new Wi-Fi network, plug it in, open the Alexa app, tap **Devices > Echo & Alexa**, then select the device you want to reconfigure. Then tap **Change** next to **Wi-Fi Network** (Figure 10-3) and follow the steps provided to connect the Echo to the Wi-Fi network where you're staying. You'll be asked to hold down the Echo's action button for about six seconds until the light ring glows orange, then connect to the Echo directly via its own temporary Wi-Fi network, then return to the Alexa app to choose an available Wi-Fi network to connect to and provide the network's password.

Once your Echo device is connected to the new W-Fi network, you'll probably want to update its location so features that require it, such as providing weather or other local information, for example, will be accurate. To update the location of an Echo device, tap **Devices > Echo & Alexa**, the name of the device, then scroll down to the **Device Location** setting.

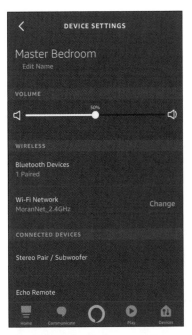

Figure 10-3: Tap **Change** to configure an Echo device on a new Wi-Fi network.

When you change the location of an Echo device, the time zone also changes automatically if applicable, and the app will remind you to review the times of any reminders or alarms set on that Echo device (as they're not automatically adjusted to reflect the new time zone).

Security Considerations When Using an Echo Device Away from Home

There are some important security considerations to be aware of whenever using an Echo device away from home. For starters, although most hotels and resorts offer public Wi-Fi access to guests, these networks aren't secure, and they're shared by lots of unknown people of unknown intent. Although communication between an Echo device and Alexa is always encrypted, since these communications involve lots of personal information to and from your Amazon account, for maximum security we recommend you never connect an

Echo to a public Wi-Fi network. Private Wi-Fi networks such as those belonging to friends and family, on the other hand, are generally okay.

If a trusted Wi-Fi network isn't available, another option is to connect your Echo to the Wi-Fi hotspot on (presumably) the same smartphone you use for the Alexa app. The process for doing this is a bit trickier than connecting to other Wi-Fi networks, though. Proceed as described in "Take an Echo Device Along when you Travel" but when asked to choose from available Wi-Fi networks to connect to, scroll down and select the **Use this device as a Wi-Fi hotspot**. You'll then be asked to type in your smartphone's Wi-Fi hotspot network name and password then activate the hotspot so the Echo can connect to it.

If you have a separate Wi-Fi hotspot device that doesn't come from the phone running the Alexa app, you can just select that Wi-Fi network from the list; you don't have to choose *Use this device as a Wi-Fi hotspot*.

If you plan to connect your Echo to a smartphone hotspot, be sure to set the device up at home first. This ensures it's running the latest software (otherwise it may not have the ability to connect to the smartphone hotspot). Also, always be mindful of the limits and costs of your mobile carrier data plan to avoid possible overage charges for excessive data use (especially if you'll be asking Alexa to stream music, for example).

Last but not least, when using an Echo device within the close confines and often thin walls of a hotel environment, consider unplugging it when you're out of the room to eliminate the possibility of picking up unintended commands. (Better yet, put it in the safe before you leave the room.)

CONCLUSION

Well, we've reached the end of the book, but your experiences with Alexa are just beginning. We've only barely scratched the surface Alexa's capabilities, but you now have a good foundation of knowledge on some of the myriad ways Alexa can enhance your daily life.

But remember, Alexa is constantly evolving and improving, so unlike most technology products you've bought in the past it doesn't begin a march to obsolescence the moment you start using it. Rather, it gets better over time; Alexa will be able to do even more for you a year from now than it can do today.

INDEX